2006 INTERNATIONAL BUILDING CODE COMPANION

About the Author

R. Dodge Woodson is a seasoned builder of as many as 60 single family homes a year, a master remodeler, a master plumber, and a master gasfitter with over 30 years of experience. Woodson opened his own business in 1979 and is the owner of The Masters Group, Inc., in Brunswick, Maine. In addition to owning and operating his contracting business, R. Dodge Woodson has taught both code and apprentice classes in the technical college system in Maine. Well known as a prolific author of many McGraw-Hill titles, Woodson's reputation and experience come together to offer readers a real-life view of professional preparation for passing the licensing exam for a trade license.

2006 INTERNATIONAL BUILDING CODE COMPANION

R. Dodge Woodson

New York Chicago San Francisco Lisbon London Madrid
Mexico City Milan New Delhi San Juan Seoul
Singapore Sydney Toronto

The *McGraw·Hill* Companies

CIP Data is on file with the Library of Congress.

Copyright © 2007 by The McGraw-Hill Companies, Inc. All rights reserved. Printed in the United States of America. Except as permitted under the United States Copyright Act of 1976, no part of this publication may be reproduced or distributed in any form or by any means, or stored in a data base or retrieval system, without the prior written permission of the publisher.

McGraw-Hill books are available at special quantity discounts to use as premiums and sales promotions, or for use in corporate training programs. For more information, please write to the Director of Special Sales, Professional Publishing, McGraw-Hill, Two Penn Plaza, New York, NY 10121-2298. Or contact your local bookstore.

1 2 3 4 5 6 7 8 9 0 DOC/DOC 0 1 3 2 1 0 9 8 7

ISBN-13: 978-0-07-148429-9
ISBN-10: 0-07-148429-9

Sponsoring Editor	**Proofreader**
Cary Sullivan	Leona Woodson
Editing Supervisor	**Production Supervisor**
David E. Fogarty	Pamela A. Pelton
Project Manager	**Composition**
Jacquie Wallace	Lone Wolf Enterprises, Ltd.
Copy Editor	**Art Director, Cover**
Wendy Lochner	Jeff Weeks
Indexer	
Roger Woodson	

This book is printed on acid-free paper.

This book is dedicated to
Adam and Afton,
the two brightest stars in my life.

CONTENTS

Acknowledgments   xvii
Introduction   xix

CHAPTER 1: ADMINISTRATION   1.1

What Does the International Building Code Include?   1.1
Building Officials   1.2
Materials   1.2
Permits   1.2
Repairs   1.3
Construction Documents   1.4
Temporary Structures and Uses   1.4
Inspections   1.5
Utility Services   1.6
Appeals   1.6
Violations   1.6
Stop-Work Order   1.6
Complying with the International Building Code   1.7

CHAPTER 2: DEFINITIONS   2.1

CHAPTER 3: USE AND OCCUPANCY CLASSIFICATION   3.1

Assembly   3.2
    Group A-1   3.2
    Group A-2   3.2
    Group A-3   3.3
    Group A-4   3.3
    Group A-5   3.4
Business   3.4
Education   3.5
Factory and Industrial   3.5
    Group F-1   3.6
    Group F-2   3.8
Hazardous Uses   3.8
Institutional Uses   3.9

Group I-3	3.11
Group I-4	3.11
Mercantile	3.12
Residential	3.12
Storage	3.13
Group S-1	3.14
Utility and Miscellaneous	3.14

CHAPTER 4: DETAILED REQUIREMENTS BASED ON USE AND OCCUPANCY

	4.1
Covered Mall Buildings	4.1
Floor Framing	4.3
Roof Framing	4.3
Floors	4.3
Roof Decks	4.3
Combustible Materials	4.3
Fire-Resistance Separations	4.4
Security	4.7
Plastic Signs	4.7
High-Rise Buildings	4.8
Atriums	4.10
Underground buildings	4.11
Motor-Vehicle-Related Occupancies	4.14
Parking Garages	4.15
Open Parking Garages	4.15
Exterior Walls	4.17
Ventilation Systems	4.18
Motor Fuel-Dispensing Facilities	4.19
Medical Facilities	4.20
Criminal Facilities	4.23
Power-Operated Sliding Doors	4.26
Vertical Openings	4.26
Motion-Picture Projection Rooms	4.28
Stages and Platforms	4.29
Stage Doors	4.30
Combustible Materials	4.31
Roof Vents	4.31
Platform Construction	4.32
Dressing Rooms	4.32
Stage Exits	4.32
Automatic Sprinkler Systems	4.32
Special Amusement Buildings	4.33
Aircraft-Related Occupancies	4.34
Airport Traffic-Control Towers	4.34
Hangers	4.35
Heliports and Helistops	4.37
Combustible Storage	4.39

Hazardous-Materials Facilities 4.40
    Emergency Alarms 4.43
    Storage 4.43
    Transport 4.44
    Excess Hazardous Materials 4.44
    Combustible Dust 4.46
    Conveyors 4.46
    Mixed Occupancies 4.46
    Dry Cleaning 4.48
    Aggregate Quantities 4.48
    Gas Detection 4.51
    Emergency Power 4.52
    Exhaust Ventilation 4.52
    Application of Flammable Liquids 4.53
    Drying Rooms 4.53
    Organic Coatings 4.53
    Occupancies 4.53
    Hydrogen Cutoff Rooms 4.54

CHAPTER 5: GENERAL BUILDING HEIGHTS AND AREAS 5.1

Definitions 5.1
General Height and Area Limitations 5.2
    Height 5.2
    Mezzanines 5.3
Equipment Platforms 5.5
    Area Modifications 5.5
        Unlimited-Area Buildings 5.7
        Mixed-Use Occupancy 5.9
        Accessory Occupancies 5.11
        Special Provisions 5.13

CHAPTER 6: TYPES OF CONSTRUCTION 6.1

Floor Framing 6.5
Roof Framing 6.5
Floors 6.6
Roof Decks 6.6
Combustible Materials 6.7

CHAPTER 7: FIRE-RESISTANCE-RATED CONSTRUCTION 7.1

Fire-Resistance Rating and Fire Tests 7.1
Exterior Walls 7.3
Shared Lots 7.4
Outside Walls 7.6
Parapets 7.9
Windows 7.9

Fire Walls   7.10
Fire Barriers and Fire Windows   7.13
Shaft Enclosures   7.14
Other Types of Shafts   7.17
Fire Partitions   7.19
Smoke Barriers and Partitions   7.20
Horizontal Assemblies   7.21
    Ceiling Panels   7.22
    Skylights   7.22
Penetrations   7.22
Fire-Resistant Joint Systems   7.24
Structural Members   7.26
    Exterior Structural Members   7.27
Opening Protectives   7.28
    Doors   7.29
    Fire Shutters   7.32
Ducts and Air Transfer Openings   7.33
    Smoke Dampers   7.34
    Fire Dampers   7.35
    Smoke Dampers   7.36
    Ceiling Radiation Dampers   7.37
Concealed Spaces   7.37
    Draft Stopping   7.39
Plaster   7.40
Thermal and Sound-Insulating Materials   7.40
Prescriptive Fire Resistance   7.41
Calculated Fire Resistance   7.41

CHAPTER 8: INTERIOR FINISHES   8.1

Definitions   8.1
Wall and Ceiling Finishes   8.2
Interior Floor Finish   8.4
Combustible Materials   8.6
Decorative Materials and Trim   8.6

CHAPTER 9: FIRE-PROTECTION SYSTEMS   9.1

Automatic Sprinkler Systems   9.2
    Group A   9.2
    Group E   9.3
    Group F   9.3
    Group H   9.4
    Group I   9.4
    Group M   9.4
    Group R   9.5
    Group S   9.5
    Type V Construction   9.8
Monitoring and Alarms   9.8

Alternative Systems 9.9
Commercial Cooking 9.10
Standpipe Systems 9.11
    Class I 9.11
    Class II and Class III 9.13
Portable Fire Extinguishers 9.14
Alarm and Detection Systems 9.14
    Group A 9.15
    Group B 9.15
    Group E 9.16
    Group F 9.16
    Group H 9.16
    Group I 9.16
Group M 9.17
Group R 9.18
    Group H 9.20
Smoke Control 9.20
Fire Command Center 9.24

CHAPTER 10: MEANS OF EGRESS 10.1

General Means of Egress 10.1
Occupant Load 10.3
Egress Width 10.5
Means-of-Egress Illumination 10.5
Accessible Means of Egress 10.7
Doors 10.10
Floors and Landings 10.13
Hardware 10.14
Gates and Turnstiles 10.15
Stairways 10.17
Ramps 10.21
Exit Signs 10.22
Handrails 10.22
Guards 10.23
Exit Access 10.24
    Aisles 10.25
Exit-Access Doorways 10.26
Floor Number Signs 10.29
Smokeproof Enclosures 10.29
Horizontal Exits 10.30
Exterior Exit Ramps and Stairways 10.30
Assembly 10.31
Emergency Escape and Rescue 10.33

CHAPTER 11: ACCESSIBILITY 11.1

Scope of Accessibility Requirements 11.1

Accessible Routes   11.3
Accessible Entrances   11.5
Parking and Passenger Loading Facilities   11.6
Dwelling and Sleeping Units   11.8
Special Occupancies   11.11
Other Features and Facilities   11.13
Platform Lifts   11.14
Signage   11.15

CHAPTER 12: INTERIOR ENVIRONMENT   12.1

Ventilation   12.1
Lighting   12.3
Yards or Courts   12.3
Sound Transmission   12.4
Interior Space Dimensions   12.4
Access to Unoccupied Spaces   12.5
Surrounding Materials   12.5

CHAPTER 13: ENERGY EFFICIENCY   13.1

CHAPTER 14: EXTERIOR WALLS   14.1

Performance Requirements   14.1
Materials   14.3
Flashing   14.4
Veneers   14.4
Window Sills   14.8
Vinyl Siding   14.9
Combustible Materials   14.9
Metal Composite Materials   14.11

CHAPTER 15: ROOF ASSEMBLIES AND ROOFTOP STRUCTURES   15.1

Performance Requirements   15.2
Fire Classification   15.3
Materials   15.5
Roof Coverings   15.5
Insulation   15.8
Structures   15.8
Reroofing   15.10

CHAPTER 16: STRUCTURAL DESIGN   16.1

Construction Documents   16.2
General Design Requirements   16.3

Load Combinations 16.5
Dead Loads 16.8
Live Loads 16.9
Snow and Wind Loads 16.14
Soil Lateral Loads 16.22
Rain Loads 16.22
Flood Loads 16.24
Earthquake Loads 16.25

CHAPTER 17: STRUCTURAL TESTS AND SPECIAL INSPECTIONS 17.1

Approvals 17.1
Special Inspections 17.2
    Steel Buildings 17.3
    Concrete Construction 17.4
    Masonry Construction 17.5
    Soils 17.5
    Pile Foundations 17.6
    Fire Resistance 17.6
    Mastic and Intumescent Coatings 17.8
    Statement of Special Inspections 17.9
    Contractor Responsibility 17.11
Design Strengths of Materials 17.11
In-Situ Load Tests 17.12
Test Standards for Joist Hangers and Connectors 17.12

CHAPTER 18: SOILS AND FOUNDATIONS 18.1

Foundation and Soil Investigations 18.1
Excavation, Grading, and Fill 18.4
Load-Bearing Values of Soils 18.6
Footings and Foundations 18.7
Dampproofing and Waterproofing 18.16
Pier and Pile Foundations 18.18
    Driven-Pile Foundations 18.22
    Micropiles 18.23
    Pier Foundations 18.24

CHAPTER 19: CONCRETE 19.1

Specifications for Tests and Materials 19.2
Durability Requirements 19.2
Concrete Quality, Mixing, and Placing 19.4
Formwork, Embedded Pipes, and Construction Joints 19.5
Structural Plain Concrete 19.6
Minimum Slab Provisions 19.6
Anchorage to Concrete 19.6
Shotcrete 19.7

Reinforced Gypsum Concrete   19.12
Concrete-Filled Pipe Columns   19.12

CHAPTER 20: ALUMINUM   20.1

CHAPTER 21: MASONRY   21.1

Masonry Construction Materials   21.1
    Type S and Type N Mortar   21.3
    AAC Masonry   21.4
Construction   21.5
Quality Assurance   21.7
Seismic Design   21.9
Empirical Design of Masonry   21.10
    Multiwythe Masonry Walls   21.12
Glass-Unit Masonry   21.16
Masonry Fireplaces   21.17

CHAPTER 22: STEEL   22.1

Identification and Protection   22.1
Structural Steel   22.1
Steel Joists   22.2
Steel Cable Structures   22.3
Steel Storage Racks   22.4
Cold-Formed Steel   22.4

CHAPTER 23: WOOD   23.1

Minimum Standards and Quality   23.1
    Fiberboard   23.2
    Trusses   23.3
    Log Buildings   23.3
    Fire-Retardant-Treated Lumber   23.4
General Construction Requirements   23.4
    Decay and Termites   23.8
General Design Requirements   23.9
Allowable Stress Design   23.14
Conventional Light-Frame Construction   23.18
    Floor Joists   23.20
    Braced Wall Lines and Panels   23.20
    Purlins   23.21
    Engineered-Wood Products   23.21
    Seismic Requirements   23.21

CHAPTER 24: GLASS AND GLAZING   24.1

General Requirements for Glass   24.1

Glass Loads   24.2
Sloped Glazing and Skylights   24.3
     Safety Glazing   24.4
Handrails and Guards   24.6
Athletic Facilities   24.7

CHAPTER 25: GYPSUM BOARD AND PLASTER   25.1

Vertical and Horizontal Assemblies   25.1
Shear-Wall Construction   25.1
Gypsum-Board Materials   25.2
Lathing and Plastering   25.3
Gypsum Construction   25.4
     Gypsum Board for Showers   25.6
Cement Plaster   25.6
Interior and Exterior Plaster   25.7
Exposed Aggregate Plaster   25.8

CHAPTER 26: PLASTIC   26.1

Foam-Plastic Insulation   26.1
Interior Finish and Trim   26.4
Plastic Veneer   26.4
Light-Transmitting Plastics   26.5
     Light-Transmitting-Plastic Wall Panels   26.6
     Light-Transmitting-Plastic Glazing   26.6
     Light-Transmitting-Plastic Roof Panels   26.8
     Light-Transmitting-Plastic Skylight Glazing   26.8
     Light-Transmitting-Plastic Interior Signs   26.9

CHAPTER 27: ELECTRICAL   27.1

CHAPTER 28: MECHANICAL SYSTEMS   28.1

CHAPTER 29: PLUMBING SYSTEMS   29.1

CHAPTER 30: ELEVATORS AND CONVEYING SYSTEMS   30.1

Hoistway Enclosures   30.1
Emergency Operations   30.2
Hoistway Venting   30.2
Conveying Systems   30.3
Machine Rooms   30.4

CHAPTER 31: SPECIAL CONSTRUCTION   31.1

Membrane Structures   31.1
Temporary Structures   31.3
Pedestrian Walkways and Tunnels   31.3
Awnings and Canopies   31.3
Marquees   31.4
Radio and Television Towers   31.4
Swimming-Pool Enclosures   31.5

CHAPTER 32: ENCROACHMENTS INTO THE
PUBLIC RIGHT-OF-WAY   32.1

CHAPTER 33: SAFEGUARDS DURING CONSTRUCTION   33.1

Construction Safeguards   33.1
Demolition   33.1
Site Work   33.2
Sanitary Provisions   33.2
Protection of Pedestrians   33.2
Exits   33.5
Standpipes   33.5
Automatic Sprinkler Systems   33.5

CHAPTER 34: EXISTING STRUCTURES   34.1

Additions, Alterations, or Repairs   34.1
Fire Escapes   34.2
Change of Occupancy   34.3
Historic Buildings   34.3
Accessibility   34.4
Compliance Alternatives   34.5

INDEX   I.1

CHAPTER 1
ADMINISTRATION

Welcome to the International Building Code administrative-procedures chapter. These next few pages will familiarize you with the laws and procedures of the 2006 International Building Codes.

This code is written to provide the minimum requirements to safeguard the public health, safety, and general welfare through structural strength, means of exit, stability, sanitation, light and ventilation, and conservation of energy. This code also applies to safety of life and property from fire and other environmental hazards and to safety of fire fighters and responders during emergency operations.

There may be sections of this code that specify different materials and methods of construction or other requirements from other parts of the code. Please note that you must follow the most restrictive rule. When there is a conflict between a general and a specific requirement, you must always apply the specific requirement. Also remember that you must pay attention to the provisions of any local, state, or federal law. Always check to make sure that you are not following one rule and breaking another.

WHAT DOES THE INTERNATIONAL BUILDING CODE INCLUDE?

The 2006 International Building Code applies to the construction, alteration, movement, enlargement, replacement, repair, equipment, use and occupancy, location, maintenance, removal, and demolition of every building or structure or any addition connected or attached to a building or structure. The exception to this rule is a detached one- or two-family dwelling or townhouse not more than three stories high and above grade plane in height with a separate exit; such structures must be built in accordance with the International Residential Code.

BUILDING OFFICIALS

Building officials are responsible for the administration and enforcement of the building code. They are chosen by the chief appointing authority of the jurisdiction. Building officials may not be held liable on a personal basis when working for a jurisdiction. Any legal suit brought against an officer because of an act performed by that officer will be defended by a legal representative of the jurisdiction.

The primary responsibility of the building official is to enforce the code. The building official also receives and reviews applications for building permits. After reviewing the construction documents, if approved, the official will issue a permit. Once a permit is issued, the building official is responsible for inspecting the premises and enforcing any requirements of the code.

The building officer is responsible for more than just permits, inspections, and code enforcement. He or she is in charge of department records. Such records include: applications, permits, inspection reports, notices, fees, and orders. Records are kept for the required retention time for public records in a jurisdiction.

MATERIALS

All materials, equipment, and devices that are approved by the building official must be constructed and installed as provided for in the approval documents. Used materials may be used if they meet the requirements for new materials. However, you cannot use old materials unless they are approved by the building official. Remember: you must seek approval from the building official to install used materials. This is also true of alternative materials, designs, and methods of construction and equipment. Approval is needed from the building official to ensure that they meet the code provisions in quality, strength, effectiveness, fire resistance, durability, and safety. Always make sure that you have valid reports from approved sources to present to the building officer when making such requests. Without this evidence the building official has the authority to require tests at your expense as evidence. It is in your best interest to have these reports available. As with other reports, they will be retained by the building official.

PERMITS

Any owner or authorized agent who intends to construct, enlarge, alter, repair, move, demolish, or change the occupancy of a building or structure is required to obtain a permit. If the alteration is an already approved electrical, gas, mechanical, or plumbing installation, the building official is authorized to issue an annual per-

rative elements, such as batting, cloth, cotton, and foam plastics. Decorative materials do not include floor coverings, ordinary window shades, or interior finishes 0.025 inch (0.64 mm) or less in thickness applied directly to a substrate.

DWELLING: A building that contains one or two units used, intended, or designed to be used, rented, leased, let out, or hired out to be occupied for living purposes.

DWELLING UNIT: A single unit providing complete, independent living facilities for one or more persons, including permanent provisions for living, sleeping, eating, cooking, and sanitation.

EXISTING STRUCTURE: A structure erected prior to the date of adoption of the current code or one for which a legal building permit has been issued.

FIRE LANE: A road or other passageway developed to allow the passage of fire vehicles.

GRADE FLOOR OPENING: A window or other opening located such that the sill height is no more than 44 inches (111.8 cm) above or below the finished ground level adjacent to the property.

HABITABLE SPACE: A space in a building for living, sleeping, eating, or cooking. Bathrooms, closets, halls, storage or utility spaces, and similar areas are not considered habitable spaces.

HISTORIC BUILDINGS: Buildings listed in or eligible for listing in the National Register of Historic Places or designated as historic under an appropriate state or local law.

JURISDICTION: The governmental unit that has adopted the building code under legislative authority.

LIGHT-FRAME CONSTRUCTION: A type of construction whose vertical and horizontal structural elements are primarily formed by a system of repetitive wood or light-gauge steel framing members.

LOT: A portion or parcel of land considered as a unit.

!Codealert

A dwelling unit is a single unit providing complete, independent living facilities for one or more persons, including permanent provisions for sleeping, eating, cooking, and sanitation.

!Codealert

A place of religious worship is a building or portion thereof intended for the performance of religious services.

LOT LINE: A line dividing one lot from another or from a street or any public place.

MARQUEE: A permanent roofed structure attached to and supported by the building and that projects into the public right-of-way.

OCCUPIABLE SPACE: A room or enclosed space that is designed for human occupancy in which individuals congregate for amusement, educational, or similar purposes or in which occupants are engaged at labor and that is equipped with means of egress and light and ventilation facilities meeting the requirements of the appropriate code.

PERMIT: An official document or certificate issued by the authority having jurisdiction that authorizes performance of a specified activity.

PERSON: An individual, heir, executor, administrator or designee, including a firm, partnership, or corporation or the agent of any of the aforesaid.

REGISTERED DESIGN PROFESSIONAL: An individual who is registered or licensed to practice his or her respective design profession as defined by the statutory requirements of the registration laws of the state of jurisdiction in which the project is to be constructed.

RELIGIOUS WORSHIP, PLACE OF: A building or portion thereof intended for the performance of religious activities.

SKYLIGHT UNIT: A factory-assembled, glazed fenestration unit containing one panel of glazing material that allows for natural lighting through an opening in the roof assembly while preserving the weather-resistant barrier of the roof.

SKYLIGHTS AND SLOPED GLAZING: Glass or other transparent or translucent glazing material installed at a slope of 15 degrees (0.26 rad) or more from vertical. Glazing materials in skylights, including unit skylights, solariums, sunrooms, roofs, and sloped walls, are included in this definition.

SLEEPING UNIT: A room or space in which people sleep; that is part of a dwelling unit that also includes permanent provisions for living, eating, and sanitation facilities.

!Codealert

Congregate living facilities with 16 or fewer occupants are permitted to comply with the construction requirements for Group R-3.

International Fire Code. There are exceptions to Group H; the following list describes uses not classified as Group H but rather in the occupancy that they most nearly resemble:

• Wholesale and retail sales and storage of flammable and combustible liquids in mercantile occupancies conforming to the International Fire Code

• Closed piping containing flammable and/or combustible liquids or gases that are used to operate machinery or equipment

• Cleaning establishments that use combustible liquid solvents with a flash point of 140 degrees F (60 degrees C) or higher in closed systems employing equipment listed by an approved testing agency, provided that this occupancy is separated from all areas of the building by 1-hour fire barriers or 1-hour horizontal assemblies or both

• Cleaning establishments that utilize a liquid solvent with a flash point at or above 200 degrees F (93 degrees C)

• Liquor stores and distributors without bulk storage

• Refrigeration systems

• Storage or utilization of materials for agricultural purposes on the premises

• Stationary batteries used for emergency power, provided that the batteries have safety venting caps and that ventilation is in accordance with the International Mechanical Code

• Corrosives not including personal or household products in their original packaging used for retail or commonly used building materials

INSTITUTIONAL USES

Institutional Group I occupancy is the use of a building or structure in which people are cared for. This can include a live-in supervised environment. People who qualify for this group have physical limitations due to health or age. Some may need medical treatment or other care. Institutional occupancies are classified as Group I-1, I-2. I-3, or I-4. Qualifications for Group I are that more than 16 people

> **!Code**alert
>
> Congregate living facilities are buildings or parts thereof that contain sleeping units where residents share bathroom and or kitchen facilities.

must be housed on a 24-hour basis. Personal care needs may be due to age, mental disability, or other reasons, but such people must be capable of responding to an emergency situation with physical assistance from staff. These residencies include but are not be limited to the following:

- Residential board and care facilities
- Assisted-living facilities
- Halfway houses
- Group homes
- Congregate-care facilities
- Social-rehabilitation facilities
- Alcohol and drug centers
- Convalescent facilities

If any of the facilities have five or fewer residents, they are classified as Group R-3 and must comply with the International Residential Code. If facilities have at least six but not more than 16 persons, they are classified as Group R-4. Group I-2 occupancy includes buildings and structures that are used for medical, psychiatric, nursing, or custodial care on a 24-hour basis for more than five persons. Care must be available on a 24-hour basis and the persons must not be capable of self-preservation. This group can include: hospitals, nursing homes (with intermediate and skilled care), mental hospitals, and detoxification facilities. A facility that has five or fewer persons is classified as Group R-3.

All facilities that provide 24-hour live-in care must comply with the International Residential Code. Such facilities do not include children under the age of 2 ½ years of age. A child-care facility that provides care to five or more children in this age bracket is classified as Group I-2. There is an exception to this classification: a facility that provides care to more than five but no more than 100 children and whose rooms are located on the same level as an exit and have a door directly to the outside are classified as Group E.

There is a lot of qualifying information that you need to refer to. Always check with the codes to make sure that your facility has the correct qualifications or you

Table 3.1 Maximum allowable quantity per control area of hazardous materials posing a physical hazard[a, j, m, n, p]

MATERIAL	CLASS	GROUP WHEN THE MAXIMUM ALLOWABLE QUANTITY IS EXCEEDED	STORAGE[b] Solid pounds (cubic feet)	STORAGE[b] Liquid gallons (pounds)	STORAGE[b] Gas (cubic feet at NTP)	USE-CLOSED SYSTEMS[b] Solid pounds (cubic feet)	USE-CLOSED SYSTEMS[b] Liquid gallons (pounds)	USE-CLOSED SYSTEMS[b] Gas (cubic feet at NTP)	USE-OPEN SYSTEMS[b] Solid pounds (cubic feet)	USE-OPEN SYSTEMS[b] Liquid gallons (pounds)
Combustible liquid[c, i]	II	H-2 or H-3	N/A	120[d, e]	N/A	N/A	120[d]	N/A	N/A	30[d]
	IIIA	H-2 or H-3		330[d, e]			330[d]			80[d]
	IIIB	N/A		13,200[e, f]			13,200[f]			3,300[f]
Combustible fiber	Loose	H-3	(100)	N/A	N/A	(100)	N/A	N/A	(20)	N/A
	baled[o]		(1,000)			(1,000)			(200)	
Consumer fireworks (Class C, Common)	1.4G	H-3	125[d, e, i]	N/A	N/A	N/A	N/A	N/A	N/A	N/A
Cryogenics flammable	N/A	H-2	N/A	45[d]	N/A	N/A	45[d]	N/A	N/A	10[d]
Cryogenics, oxidizing	N/A	H-3	N/A	45[d]	N/A	N/A	45[d]	N/A	N/A	10[d]
Explosives	Division 1.1	H-1	1[e, g]	(1)[e, g]	N/A	0.25[e]	(0.25)[e, g]	N/A	0.25[e]	(0.25)[e, g]
	Division 1.2	H-1	1[e, g]	(1)[e, g]	N/A	0.25[e, f]	(0.25)[e, g]	N/A	0.25[e]	(0.25)[e, g]
	Division 1.3	H-1 or 2	5[e, g]	(5)[e, g]	N/A	1[e]	(1)[e, g]	N/A	1[e]	(1)[e, g]
	Division 1.4	H-3	50[e, g]	(50)[e, g]	N/A	50[e]	(50)[e, g]	N/A	N/A	N/A
	Division 1.4G	H-3	N/A	N/A	N/A	N/A	N/A	N/A	N/A	N/A
	Division 1.5	H-1	1[e, g]	(1)[e, g]	N/A	0.25[e]	(0.25)[e, g]	N/A	0.25[e]	(0.25)[e, g]
	Division 1.6	H-1	1[d, e, g]	N/A	N/A	N/A	N/A	N/A	N/A	N/A
Flammable gas	Gaseous	H-2	N/A	N/A	1,000[d, e]	N/A	N/A	1,000[d, e]	N/A	N/A
	liquefied			30[d, e]	N/A		30[d, e]	N/A		
Flammable liquid[e]	1A	H-2	N/A	30[d, e]	N/A	N/A	30[d]	N/A	N/A	10[d]
	1B and 1C	H-2 or H-3		120[d, e]			120[d]			30[d]
Combination flammable liquid (1A, 1B, 1C)	N/A	H-2 or H-3	N/A	120[d, e, h]	N/A	N/A	120[d, h]	N/A	N/A	30[d, h]
Flammable solid	N/A	H-3	125[d, e]	N/A	N/A	125[d]	N/A	N/A	25[d]	N/A
Organic peroxide	UD	H-1	1[e, g]	(1)[e, g]	N/A	0.25[e, g]	(0.25)[e, g]	N/A	0.25[e, g]	(0.25)[e, g]
	I	H-2	5[d, e]	(5)[d, e]	N/A	1[d]	(1)	N/A	1[d]	(1)[d]
	II	H-3	50[d, e]	(50)[d, e]	N/A	50[d]	(50)[d]	N/A	10[d]	(10)[d]
	III	H-3	125[d, e]	(125)[d, e]	N/A	125[d]	(125)[d]	N/A	25[d]	(25)[d]
	IV	N/A	NL	NL	N/A	NL	NL	N/A	NL	NL
	V	N/A	NL	NL	N/A	NL	NL	N/A	NL	NL
Oxidizer	4	H-1	1[e, g]	(1)[e, g]	N/A	0.25[e]	(0.25)[e, g]	N/A	0.25[e, g]	(0.25)[e, g]
	3[k]	H-2 or H-3	10[d, e]	(10)[d, e]	N/A	2[d]	(2)[d]	N/A	2[d]	(2)[d]
	2	H-3	250[d, e]	(250)[d, e]	N/A	250[d]	(250)[d]	N/A	50[d]	(50)[d]
	1	N/A	4,000[d, e, f]	(4,000)[d, e, f]	N/A	4,000[d]	(4,000)[f]	N/A	1,000[d]	(1,000)[f]
Oxidizing gas	Gaseous	H-3	N/A	N/A	1,500[d, e]	N/A	N/A	1,500[d, e]	N/A	N/A
	liquefied		N/A	15[d, e]	N/A	N/A	15[d, e]	N/A	N/A	N/A

(continued)

TABLE 3.1 Maximum allowable quantity per control area of hazardous materials posing a physical hazard[a, j, m, n, p] *(continued)*

MATERIAL	CLASS	GROUP WHEN THE MAXIMUM ALLOWABLE QUANTITY IS EXCEEDED	STORAGE[b]			USE-CLOSED SYSTEMS[b]			USE-OPEN SYSTEMS[b]	
			Solid pounds (cubic feet)	Liquid gallons (pounds)	Gas (cubic feet at NTP)	Solid pounds (cubic feet)	Liquid gallons (pounds)	Gas (cubic feet at NTP)	Solid pounds (cubic feet)	Liquid gallons (pounds)
Pyrophoric material	N/A	H-2	4[e-g]	(4)[e-g]	50[e-g]	1[g]	(1)[g]	10[e-g]	0	0
Unstable (reactive)	4	H-1	1[g]	(1)[g]	10[d,e]	0.25[g]	(0.25)[g]	2[e-g]	0.25[g]	(0.25)[g]
	3	H-1 or H-2	5[d,e]	(5)[d,e]	50[d,e]	1[d]	(1)	10[d,e]	1[d]	(1)[d]
	2	H-3	50[d,e]	(50)[d,e]	250[d,e]	50[d]	(50)[d]	250[d,e]	10[d]	(10)[d]
	1	N/A	NL	NL	N/A	NL	NL	N/A	NL	NL
Water reactive	3	H-2	5[d,e]	(5)[d,e]	N/A	5[d]	(5)[d]	N/A	1[d]	(1)[d]
	2	H-3	50[d,e]	(50)[d,e]	N/A	50[d]	(50)[d]	N/A	10[d]	(10)[d]
	1	N/A	NL	NL	N/A	NL	NL	N/A	NL	NL

For SI: 1 cubic foot = 0.023 m³, 1 pound = 0.454 kg, 1 gallon = 3.785 L.

NL = Not Limited; N/A = Not Applicable; UD = Unclassified Detonable

a. For use of control areas, see Section 414.2.

b. The aggregate quantity in use and storage shall not exceed the quantity listed for storage.

c. The quantities of alcoholic beverages in retail and wholesale sales occupancies shall not be limited providing the liquids are packaged in individual containers not exceeding 1.3 gallons. In retail and wholesale sales occupancies, the quantities of medicines, foodstuffs, consumer or industrial products, and cosmetics containing not more than 50 percent by volume of water-miscible liquids with the remainder of the solutions not being flammable, shall not be limited, provided that such materials are packaged in individual containers not exceeding 1.3 gallons.

d. Maximum allowable quantities shall be increased 100 percent in buildings equipped throughout with an automatic sprinkler system in accordance with Section 903.3.1.1. Where Note e also applies, the increase for both notes shall be applied accumulatively.

e. Maximum allowable quantities shall be increased 100 percent when stored in approved storage cabinets, day boxes, gas cabinets, exhausted enclosures or safety cans. Where Note d also applies, the increase for both notes shall be applied accumulatively.

f. The permitted quantities shall not be limited in a building equipped throughout with an automatic sprinkler system in accordance with Section 903.3.1.1.

g. Permitted only in buildings equipped throughout with an automatic sprinkler system in accordance with Section 903.3.1.1.

h. Containing not more than the maximum allowable quantity per control area of Class IA, IB or IC flammable liquids.

i. Inside a building, the maximum capacity of a combustible liquid storage system that is connected to a fuel-oil piping system shall be 660 gallons provided such system complies with the *International Fire Code.*

j. Quantities in parenthesis indicate quantity units in parenthesis at the head of each column.

k. A maximum quantity of 200 pounds of solid or 20 gallons of liquid Class 3 oxidizers is allowed when such materials are necessary for maintenance purposes, operation or sanitation of equipment. Storage containers and the manner of storage shall be approved.

l. Net weight of the pyrotechnic composition of the fireworks. Where the net weight of the pyrotechnic composition of the fireworks is not known, 25 percent of the gross weight of the fireworks, including packaging, shall be used.

m. For gallons of liquids, divide the amount in pounds by 10 in accordance with Section 2703.1.2 of the *International Fire Code.*

n. For storage and display quantities in Group M and storage quantities in Group S occupancies complying with Section 414.2.4, see Tables 414.2.5(1) and 414.2.5(2).

o. Densely packed baled cotton that complies with the packing requirements of ISO 8115 shall not be included in this material class.

p. The following shall not be included in determining the maximum allowable quantities:
   1. Liquid or gaseous fuel in fuel tanks on vehicles.
   2. Liquid or gaseous fuel in fuel tanks on motorized equipment operated in accordance with this code.
   3. Gaseous fuels in piping systems and fixed appliances regulated by the *International Fuel Gas Code.*
   4. Liquid fuels in piping systems and fixed appliances regulated by the *International Mechanical Code.*

As you can see, there are many occupancy uses and exceptions that you must be aware of when constructing your building or structure. Read the code book, follow and implement the provisions, and if you have questions, don't hesitate to contact your building official.

TABLE 4.1 Fire-resistance rating requirements for exterior walls based on fire separation distance[ae].

FIRE SEPARATION DISTANCE $=X$ (feet)	TYPE OF CONSTRUCTION	OCCUPANCY GROUP H	OCCUPANCY GROUP F-1, M, S-1	OCCUPANCY GROUP A, B, E, F-2, I, R, S-2, U[b]
$X < 5^c$	All	3	2	1
$5 \leqslant X < 10$	IA	3	2	1
	Others	2	1	1
$10 \leqslant X < 30$	IA, IB	2	1	1[d]
	IIB, VB	1	0	0
	Others	1	1	1[d]
$X \geqslant 30$	All	0	0	0

For SI: 1 foot = 304.8 mm.
a. Load-bearing exterior walls shall also comply with the fire-resistance rating requirements of Table 601.
b. For special requirements for Group U occupancies see Section 406.1.2.
c. See Section 705.1.1 for party walls.
d. Open parking garages complying with Section 406 shall not be required to have a fire-resistance rating.
e. The fire-resistance rating of an exterior wall is determined based upon the fire separation distance of the exterior wall and the story in which the wall is located.

is separate from the tenant spaces or anchors. If tenant spaces use the same sprinkler system, they must have separate individual controls. However, the 2006 International Building Code does not require the use of an automatic sprinkler system in spaces or areas of open parking garages.

The covered mall building must be equipped throughout with a standpipe system. The code requires more than just fire-resistant walls and sprinkler systems for different areas of a covered mall building. For example, kiosks or other similar structures (both temporary and permanent) cannot be located within the mall unless constructed of fire-retardant wood that complies with the code, foam plastics that have a maximum heat-release rate not greater than 100kW (105 Btu/h) when tested in accordance with the exhibit booth protocol in UL 1975, or aluminum composite material (ACM) having a flame-spread index of not more than 25 and a smoke-developed index of not more than 450 when tested as an assembly in the maximum thickness intended for use in accordance with ASTM E 84. All kiosks or similar structures located within the mall must also be provided with approved fire-suppression and detection devices. Spacing within kiosks or similar structures must be at least 20 feet (6.1 m), and they must have a maximum area of 300 square feet (28 m²).

There are many structures that require special detailed requirements besides mall buildings or garages. We all want our children to be safe, and we don't want to have to worry about the structures that they play on at a mall. We want to know that these structures are made of only the safest materials. Thanks to the 2006 International Building Code we can be assured that the playground structures that our children play on are built to code. Structures that are intended for children's playgrounds that exceed 10 feet (3.5 m) in height and 150 square feet (14 m²) must comply with the following requirements:

- Fire-retardant-treated wood

- Light-transmitting plastics

- Foam plastics (including the pipe foam used in soft-contained play equipment structures) having a maximum heat-release rate not greater than 100kW when tested in accordance with UL 1975

- Aluminum composite material (ACM) meeting the requirements of Class A interior finish in accordance with Chapter 8 of the code when tested as an assembly in the maximum thickness

- Textiles and films complying with the flame performance criteria in NFPA 701

- Plastic materials used to construct rigid components of soft-contained play equipment structures (such as tubes, windows, panels, and decks) meeting the UL 94 V-2 classification when tested in accordance with UL 94

- Foam plastics must be covered by a fabric, coating, or film that meets the flame performance criteria of NFPA 701

- The interior floor under the children's playground structure must be a Class I interior floor finish when tested in accordance with NFPA 253

Any playground structure for the use of children located within the mall must be provided with the same level of approved fire-suppression and -detection devices required for kiosks and similar structures. Children's playground structures are also required, just as kiosks are, to be separated from other structures. Playground structures are required to have a minimum horizontal separation of 20 feet (6.1 mm) from any other structures within the mall. They also cannot exceed 300 square feet (28 m²) unless a special investigation has demonstrated adequate fire safety.

!Codealert

Where exit passageways provide a secondary means of egress from a tenant space, doorways to the exit passageway shall be protected by 1-hour fire-door assemblies that are self- or automatic-closing by smoke detection.

Security

One of the most important safety features of covered malls are security grilles and doors. Horizontal sliding or vertical security grilles or doors that are part of a required means of exit have to conform to rules as well. These rules are:

- They must remain in the full open position during the time periods that the mall is occupied by the general public.

- Doors or grilles cannot be closed when there are 10 or more persons occupying a space that has only one exit or when there are 50 or more persons occupying a space that has more than one exit.

- The doors or grilles must be able to be open from the inside of the building without the use of any special knowledge or special efforts where the space is occupied.

- Where two or more exits are required, not more than one-half of the exits are permitted to include a horizontal sliding or vertical rolling grille or door.

It is mandatory that covered mall buildings that exceed 50,000 square feet (4645 m^2) be provided with standby power systems that are capable of operating the emergency communication system. These emergency voice/alarm communication systems must be accessible to the fire department.

Plastic Signs

Plastic signs that are applied to the storefront of any tenant space facing the mall must adhere to the following rules:

- Plastic signs must not exceed 20 percent of the wall area facing the mall.

- Horizontal plastic signs must not exceed a height of 36 inches (914 mm), and vertical signs cannot be higher than 96 inches (2438 mm); the width cannot be wider than 36 inches (914 mm).

- Plastic signs must be located a minimum distance of at least 18 inches (457 mm) from adjacent tenants.

- Plastics other than foam used in signs must be light-transmitting plastics and shall have a self-ignition temperature of 650°F (343°C) or greater when tested in accordance with ASTM D 1929 and a flame spread index not greater than 75 and a smoke-developed index not greater than 450 when tested in the manner intended for use in accordance with ASTM E 84.

- Edges and backs of plastic signs must be fully encased in metal.

- Foam plastics used in signs must have a maximum heat-release rate of 150 kilowatts when tested in accordance with UL 1975, and they must have the physical characteristics specified in this section. Foam plastics used in signs installed in accordance with this section are not required to comply with the flame spread and smoke-developed indexes specified in the plastics section of the code.

- The minimum density of foam plastics used in signs must not be less than 20 pounds per cubic feet (pcf) (320kg/m^3), and the thickness must not be greater than ½ inch (12.7 mm).

This concludes the section on covered mall buildings and similar structures. Remember that buildings, structures, and rooms that contain controls for air-conditioning systems, automatic fire-extinguishing systems, or other detection or control elements must be identified for use by the fire department. You don't want to have wasted all your time, money, and efforts installing these systems by failing to make the fire department aware of the controls and where they are located. As always, when any questions arise seek the advice of a professional or a building official.

HIGH-RISE BUILDINGS

The provisions of this section apply to buildings with an occupied floor located more than 75 feet above the lowest level of fire-department-vehicle access. There are five exceptions to these provisions:

• Airport traffic-control towers

• Open parking spaces

• Buildings with occupancy in Group A-5

• Low-hazard special industrial occupancies

• Building with occupancy in Group H-1, H-2, or H-3

Buildings and structures must be equipped throughout with an automatic sprinkler system and a secondary water supply where required. We know from earlier discussion that open parking garages are an exception to this provision. Reduction in the fire-resistant rating is allowed in buildings that have sprinkler control valves equipped with supervisory initiating devices and water-flow initiating devices for each floor. For buildings not greater than 420 feet (128 m) in height, Type IA construction is allowed to be reduced to Type IB with the exception of columns supporting the floor, which may not be reduced.

In other groups not found in Table 4.2 such as Groups F-1, M, and S-1, Type IB construction is allowed to be reduced to Type IIA. The height and area limitations of the reduced construction type ares allowed to be the same for as the original construction type.

For buildings not greater than 420 feet (128 m) in height, the required fire-resistance rating of the fire barriers enclosing vertical shafts, other than exit enclosures and elevator hoistway enclosures, must be reduced to 1 hour where automatic sprinklers are installed within the shafts at the top and at alternate floor levels. Emergency escape and rescue openings required by Chapter 10 of the code are not required for this type of construction. Smoke detectors, emergency voice/alarm, and fire-department communication systems must be installed and provided for emergency use in a location approved by the fire department. I spoke

When compartmentation is required, each compartment needs to have an independent smoke-control system. It must be automatically activated and capable of manual operation. Underground buildings require both fire-alarm systems and a public-address system where required in Chapter 9. Each floor of an underground building must have a minimum of two exits; if compartmentation is required, the compartments must have a minimum of one exit and an exit access doorway into an adjoining compartment.

Staircases in underground buildings that serve floor levels of more than 30 feet (9.1 m) below their level of exit must comply with the requirements for a smokeproof enclosure. This is described in Chapter 10 of the code. It is important that your underground building have standby power emergency power for blackouts, storms, or other situations that cause power outages. Having these systems in place can save lives and make exiting the building easier. These loads classify as standby-power loads:

• Smoke control systems

• Ventilation and automatic fire-detection equipment for smokeproof enclosures

• Fire pumps

Remember that your standby power must be provided for elevators as well in accordance with Chapter 3. The standby power system must start up within 60 seconds after the failure of normal power supply. The following qualify as emergency power loads:

• Emergency voice/alarm communication systems

• Fire-alarm systems

• Automatic fire-detection systems

• Elevator-car lighting

• Means of exit and exit-sign illumination as required by Chapter 10

You must have a standpipe system equipped throughout the underground building in accordance with Chapter 9.

!Definitionalert

Open Parking Garage: A structure or portion of a structure with openings on two or more sides that is used for the parking or storage of private motor vehicles.

MOTOR-VEHICLE-RELATED OCCUPANCIES

This section relates to private garages and carports. These are classified as Group U. For your building to classify as Group U, no repair work or dispensing of fuel can be provided. It is strictly for the storage of private or pleasure-type motor vehicles. This type of building is permitted to be 3,000 square feet (279 m^2) if the following provisions are met. For a mixed- occupancy (remember that some members of different groups may qualify for more than one group) use, the exterior wall and opening protection for the Group U portion of the building is also required for the major occupancy of the building. For such a mixed-occupancy building, the allowable floor area must be designated for the major occupancy type. If the building contains only a Group U occupancy, the exterior wall is not required to have a fire-resistance rating, and the area of openings will not be limited when the fire-separation distance is 5 feet (1.5 cm) or more. More than one 3,000-square-foot Group U occupancy is permitted in the same building if each area is separated by firewalls complying with Chapter 7.

The provisions for a carport are different than those of a garage. Carports must be open on at least two sides, and the floor surfaces must be of an approved noncombustible material.

Asphalt surfaces may be permitted at ground levels in carports. The area of the floor used to park automobiles or other vehicles must be sloped for liquids to enter the drain or the main doorway. If your facility is not open on at least two sides, then it is a garage and must comply with the provisions of this section for garages.

A private garage must be kept separate from the dwelling unit and its attic area by a minimum ½-inch (12.7 mm) gypsum board applied to the garage side. If your garage is under living space, it needs to be separated by no less than a 5/8-inch Type X gypsum board or equivalent. The openings between a private garage and the home must have either sold wood doors or solid or honeycomb-core steel doors in compliance with Chapter 7.

Doors from a private garage that enter directly into a room used as a bedroom or sleeping area are not permitted. All doors must be self-closing and self-latching. Any ducts in the walls and ceiling of a private garage that separate the home from the garage must be constructed of a minimum of 0.019-inch (0.48 mm) sheet

!Codealert

Smoke control is not required for atriums that connect only two stories.

steel with no openings into the garage. A separation such as that above is not needed between a Group R-3 and U carport provided the carport is entirely open on two or more sides and there are not closed areas above.

Parking Garages

Parking garages, which are considered to be different from personal garages, are classified as either open or enclosed. See Chapter 5 for special provisions for parking garages with regard to classification.

Each floor level in a parking garage that is for vehicle and pedestrian traffic must have a height no less than 7 feet (2.1 m). Vehicle and pedestrian areas that accommodate van-accessible parking, as required by Chapter 11, must conform to ICC A117.1. It is mandatory that guards be provided in accordance with Chapter 10. These guards will be placed at exterior and interior vertical openings on floor and roof areas where vehicles are parked or moved and where the vertical distance to the ground or surface directly below exceeds 30 inches (7.6 m).

The parking areas of the garage will have exterior or interior walls or vehicle barriers, except at pedestrian or vehicular accesses, and must be designed in accordance with Chapter 16. Barriers not less than 2 feet high must be placed at the end of drive lanes and at the end of parking spaces where the difference in adjacent floor elevation is greater than 1 foot (.3 m). The only exception to this are vehicle storage compartments in a mechanical-access parking garage. Any ramps used for vertical circulation or for parking must not exceed a slope of 1:15 (6.67 percent), and vehicle ramps cannot be considered as required ramps unless pedestrian facilities are provided. Parking garages are separated from other occupancies in accordance with Chapter 5.

Parking garages that are connected to any room in which there is a fuel-fired appliance must be separated by means of a vestibule that provides a two-doorway separation, with the exception that a single door will be allowed provided that the sources of ignition in the appliance are at least 18 inches (45.7 cm) above the floor. Parking garages, like personal garages, cannot have openings directly into a room used for sleeping.

Open Parking Garages

Open parking garages must be of Type I, II, or IV construction. Open parking garages must meet the design requirements of Chapter 16. For natural-ventilation purposes, the outside area of the building or structure must have openings on two or more sides and must be evenly distributed. The area of these openings in outside walls on a tier must be at least 20 percent of the total perimeter-wall area of each tier.

The aggregated lengths of the openings are considered to provide natural ventilation but must occupy a minimum of 40 percent of the perimeter of the tier. If the

TABLE 4.3 Open parking garages area and height.

TYPE OF CONSTRUCTION	AREA PER TIER (square feet)	HEIGHT (in tiers)		
			Mechanical access	
			Automatic sprinkler system	
		Ramp access	No	Yes
IA	Unlimited	Unlimited	Unlimited	Unlimited
IB	Unlimited	12 tiers	12 tiers	18 tiers
IIA	50,000	10 tiers	10 tiers	15 tiers
IIB	50,000	8 tiers	8 tiers	12 tiers
IV	50,000	4 tiers	4 tiers	4 tiers

For SI: 1 square foot = 0.0929 m^2.

required openings are evenly distributed over two opposing sides of the building, the openings are not required to be distributed over 40 percent of the building.

Mixed uses are allowed in the same building as an open parking garage provided that they meet the code provisions in Chapter 5. The area and height of open parking garages are also limited to the provisions in Chapter 5. If the open garage is used exclusively for the parking or storage of private motor vehicles, with no other uses in the building, the area and height will be permitted to comply with Table 4.3 and this section.

An exception to this rule is that the grade-level tier is permitted to contain an office, waiting area, and toilet rooms having a total combined area of not more than 1,000 square feet (93 m^2). This area does not need to be separated from the open parking garage.

In parking garages that are open and have a spiral or sloping floor, the horizontal projection of the structure at any cross section must not exceed the allowable area per parking tier. For an open parking garage with a continuous spiral floor, each 9 feet 6 inches (2.9 m) of height will be considered a tier. The clear height of a parking tier must not be less than 7 feet (2.1 m), except that a lower clear height is allowed in mechanical-access open parking garages but only when approved by the building official.

Garages with sides that are open on three-fourths of the building's perimeter can be increased by 25 percent in area and one tier in height. Garages with open sides all around the entire building's perimeter are permitted to be increased by 50 percent in area and one tier in height. For your garage to be considered open under these provisions, the total area of the openings along the side are no less than 50 percent of the interior area of the side at each tier. And these openings must be equally distributed along the length of the tier. The allowable tier areas in Table 4.3 above must be increased for open parking garages constructed to heights less than the maximum shown in the table. The gross tier area of the garage cannot exceed that allowed for the higher structure. At least three sides of each larger tier must have continuous horizontal openings no less than 30 inches (7.6 cm) in clear height. These must extend for a least 80 percent of the length of the sides, and no part of the larger tier must be more than 200 feet (61 m) horizontally from the opening. In addition, each opening must face a street or a yard accessible to a street with a width of at least 30 feet (9.1 m) for the full length of the opening, and standpipes must be provided in each tier.

The code rules for open parking garages that are of Type II construction, with all sides open, are unlimited in allowable area where the height does not exceed 75 feet (22.9 m). To be considered open, the total area of openings along the side cannot be less than 50 percent of the interior area of the side at each tier. Such openings must be equally distributed along the length of the tier as well. All portions of these tiers must be within 200 feet (60.9 m) horizontally from openings or other natural ventilation as defined earlier in this chapter. These openings are permitted to be provided in courts with a minimum width of 30 feet (9.1 m) for the full width of the openings.

You can see that there are many similarities in the 2006 International Building Code book regarding covered malls, atriums, and garages, both open and covered. It is up to you to be sure of the codes you are applying to your building(s) or structure(s). A simple misunderstanding or misinterpretation can be detrimental to your building inspection. Please read, and reread if you need to, to be sure you understand each section of the code that applies to you to ensure that you receive a positive building inspection.

Exterior Walls

Exterior walls and openings in exterior walls must comply with Tables 4.1 and 4.2. The distance to an adjacent lot line will be determined in accordance with

!Codealert

Vehicle ramps that are utilized for vertical circulation as well as for parking shall not exceed a slope of 6.67 percent.

Table 4.2 and Chapter 7. All buildings or structures require exits; obviously garages do as well. Where persons other than parking attendants are allowed, open parking garages must meet the exit requirements of Chapter 10 and must not be less than two 36-inch-wide (9.1 cm) exit stairways. Elevators are permitted to be installed for the use of employees only, provided they are completely enclosed by noncombustible materials.

Where required by other provisions of this code, automatic sprinkler systems and standpipes must be installed in accordance with the provisions of Chapter 9. The following uses and alterations are not permitted:

• Vehicle repair work

• Parking of buses, trucks, and similar vehicles

• Partial or complete closing of required openings in exterior walls by tarps or any other means

• Dispensing of fuel

Enclosed vehicle parking garages and portions of such garages that do not meet the definition of open parking garages must be limited to the allowable heights and areas specified. Chapter 5 has special provisions for parking garages. Roof parking is permitted for enclosed vehicle parking garages.

Ventilation Systems

A mechanical ventilation system must be provided in accordance with the International Mechanical Code. Facilities that dispense motor fuel must be constructed in accordance with the International Fire Code and Chapter 4 of the International Building Code. Canopies that cover fuels must have a clear, unobstructed height no less than 13 feet 6 inches (4.1 m) to the lowest projecting element in the automobile drive-through area. Canopies and the supports that cover the pumps must be constructed of noncombustible materials, fire-retardant-treated wood complying with Chapter 23, and wood of Type IV sizes or of construction providing 1-hour fire resistance. Combustible materials that are used in or on a canopy must comply with one of the following:

• Shielded from the pumps by a noncombustible element of the canopy or wood of Type IV sizes

• Plastics covered by aluminum facing having a minimum thickness of 0.010 inch (.3 mm) or corrosion-resistant steel having a minimum base metal thickness of 0.016 inch (0.41 mm); the plastic must have a flame-spread index of 25 or less; a smoke-developed index of 450 or less when tested in the form intended for use in accordance with ASTM E 84; and self-ignition temperature of 650°F (343°C) or greater when tested in accordance with ASTM D 1929

• Panels constructed of light-transmitting plastic materials can be installed in canopies erected over motor-vehicle fuel-dispensing stations , provided the panels are located at least 10 feet (3 m) from any building on the same lot and face

fined to a bed, at least 6 net square feet (0.56 m^2) per occupant must be provided on each side of each smoke barrier for the total number of occupants in adjoining smoke compartments.

All smoke compartments have to provide an independent means of exit so that staff, residents, patients, and family do not have to return through the smoke compartment to get to safety. Automatic sprinkler systems must be installed in smoke compartments that contain patient sleeping areas. The automatic sprinkler systems must be constructed in accordance with the provisions in Chapter 9. The smoke compartments must also be equipped with approved quick-response or residential sprinklers in accordance with the provisions in Chapter 9 as well.

Automatic fire detection for corridors in nursing homes (both intermediate and skilled nursing facilities), detoxification facilities, and spaces permitted to be open to the corridors as an independent exit must also be equipped with an automatic fire-detection system. Hospitals, too, must have smoke detection installed as required. There are two exceptions:

- Corridor smoke detection is not required where patient sleeping units are provided with smoke detectors that comply with UL 268. These detectors must provide a visual display on the corridor side of each patient sleeping unit and an audible and visual alarm at the nursing station that monitors each unit.
- Corridor smoke detection is not required where patient sleeping-unit doors are equipped with automatic door-closing devices with integral smoke detectors on the inside of the units. These must be installed in accordance with their listing, provided that the detectors perform the required alerting function.

Some facilities have secured yards installed for the safety of the patients and staff. Grounds are permitted to be fenced, and gates are permitted to be equipped with locks provided that safe dispersal areas having 30 net square feet (2.8 m^2) for bed and litter patients and 6 net square feet (0.56 m^2) for patients who can walk and other occupants are located between the building and the fence. These safe dispersal areas cannot be located less than 50 feet (15.2 m) from the building that they are used for.

CRIMINAL FACILITIES

Not all facilities that restrict people for various reasons are medical. I-3 buildings and structures house criminal offenders. Where security operations require the locking of means of exit, provisions must be made for the release of occupants at all times. Exits that pass through other areas of use must, at a minimum, conform to requirements for detention and correctional occupancies. It is possible to exit through a horizontal exit into other adjoining occupancy exit provisions but that do comply with requirements used in the appropriate occupancy, as long as the occupancy is not a high-hazard use. The means of exit must comply with Chapter 10 of the code except as modified or as provided for in this section. Doors to resident sleeping areas need to have clear width of no less than 28 inches (71 cm). If the door used for an exit is a sliding door, the force to slide the door fully open cannot exceed 50 pounds (220 N) with an upright force against the door of 50 pounds (220 N).

Facilities that have spiral stairs that conform to the requirements of Chapter 10 are permitted for access to and between staff locations. Facilities may have exit doors that lead into a fenced or walled courtyard; these courtyards must be of a size that accommodates all occupants with a minimum of 50 feet (15.2 m) from the building with a net area of 15 square feet (1.4 m^2) per person. During an emergency at a correctional facility a sally port may be used as a means of exit when there are provisions for continuous and unobstructed passage through the port during an emergency exit condition. Of all the required exit enclosures in each building, one of them may have glazing installed in each door and inside wall at each landing level, providing access to the enclosure, only if the following conditions are met:

• The exit enclosure cannot serve more than four floor levels.

• Exit doors cannot be less than ¾-hour fire-door assemblies complying with the provisions in Chapter 7.

• The total area of glazing at each floor level cannot exceed 5,000 square inches (3 m^2), and individual panels of glazing cannot exceed 1,296 square inches (0.84 m^2).

• The glazing will be protected on both sides by an automatic fire sprinkler system. The sprinkler system will be designed to completely soak the entire surface of any glazing affected by fire.

• The glazing must be in a gasketed frame and installed in such a manner that the framing system will deflect without breaking the glass before the sprinkler system turns on.

!Codealert

When dealing with facilities for hazardous materials, walls shall not obstruct more than one side of a structure, except that walls shall be permitted to obstruct portions of multiple sides of the structure, provided that the obstructed area does not exceed 25 percent of the structure's perimeter.

You must have standby or emergency power should an emergency or crisis occur. Where mechanical ventilation, treatment systems, temperature control, alarm, detection, or other electrically operated systems are required, such systems must be provided with an emergency or standby power system in accordance with the IBC, code or the ICC Electrical Code. The following are exceptions:

• Storage areas for Class I and II oxidizers

• Storage areas for Class III, IV and V organic peroxides

• Storage, use, and handling areas for highly toxic or toxic materials as provided for in the International Fire Code

• Standby power for mechanical ventilation, treatment systems, and temperature-control systems are not required where an approved fail-safe engineered system is installed

Rooms, buildings, or areas used for the storage of solid and liquid hazardous materials must be provided with a means to control spillage and to contain or drain off the spillage and fire-protection water discharged in the storage area where required in the International Fire Code; this includes the methods of spill control and outdoor storage, dispensing, and use of hazardous materials.

Protection from weather is a very important issue in regard to hazardous materials. Where weather protection is provided for sheltering outdoor hazardous-material storage or use areas, such areas must be considered outdoor storage or use and comply with the section provisions.

Walls must not obstruct more than one side of the structure unless the obstructed area does not exceed 25 percent of the structure's perimeter. The distance from the structure to buildings, lot lines, public ways, or means of exits cannot be less than the distance required for an outside hazardous-material storage or use area without weather protection. The overhead structure must be of approved noncombustible construction with a maximum area of 1,500 square feet ($140\ m^2$). The increases allowed by Chapter 5 are permitted.

Emergency Alarms

Emergency alarms for the detection and notification of an emergency condition in Group H occupancies must be provided.

Storage

An approved manual emergency system must be provided in buildings, rooms, or areas used for storage of hazardous materials. Emergency alarm-initiating devices must be installed outside each interior exit or exit access door of storage buildings, rooms, or areas. The activation of an alarm-initiating device must sound a local alarm to alert occupants of an emergency situation involving hazardous materials.

Transport

Where hazardous materials having a hazard ranking of 3 or 4 in accordance with NFPA 704 are transported through corridors or exit enclosures, there must be an emergency telephone system, a local manual alarm station, or an approved alarm-initiating device at not more than 150-foot (45.7 m) intervals and at each exit and exit access doorway throughout the transport route. The signal must be relayed to an approved central, proprietary, or remote station service or to a location that is constantly attended or supervised; this must also initiate a local audible alarm.

This next section covers Groups H-1, H-2, H-3, H-4 and H-5. Unless you have these groups memorized, you will probably need to refer to Chapter 3 of this book, which defines each of the categories of these groups.

Excess Hazardous Materials

There are several groups pertaining to hazardous materials. The provisions of this section apply to the storage and use of hazardous materials in excess of the maximum allowable quantities per control area listed in Chapter 3, Groups H-1, H-2, H-3, H-4, and H-5. Buildings and structures with occupancy in Group H must also comply with this section and the International Fire Code.

Group H occupancies must be located on property in accordance with other provisions of this chapter. In Groups H-2 and H-3, not less than 25 percent of the perimeter wall of the occupancy must be an external wall. The following are exceptions:

• Liquid use, dispensing, and mixing rooms having a floor area not more than 500 square feet (46.5 m²) need not be located on the outer perimeter of the building in accordance with the International Fire Code and NFPA 30.

• Liquid storage rooms having a floor area not more than 1,000 square feet (93 m²) need not be located on the outer perimeter in accordance with the International Fire Code and NFPA 30.

• Spray-paint booths that comply with the International Fire Code need not be located on the outer perimeter.

Regardless of any other provisions, buildings containing Group H occupancies must be set back to the minimum fire separation distance as discussed below. Distances must be measured from the walls enclosing the occupancy to lot lines, including those on public property. You may not use distances to assumed lot lines to establish the minimum fire separation for buildings on sites where explosives are manufactured or used when separation is provided in accordance with the quantity distance tables specified for explosive materials in the International Fire Code.

Group H-1 cannot be less than 75 feet (22.9 m) and not less than required by the International Fire Code except for fireworks-manufacturing buildings separated in accordance with NFPA 1124; buildings containing organic peroxides; and unclassified detonatable, unstable reactive materials of class 3 and 4 (detonatable

manual alarm station, or another approved alarm-initiating device within the corridors at no more than 150-foot (45.7 m) intervals. Self-closing doors must have a fire-protection rating of not less than 1 hour.

Service corridors are of a different nature, and the coding is somewhat different. They are classified as Group H-5, and service corridors are separate from corridors as previously discussed in this chapter. Service corridors cannot be used as a required corridor, and they must be mechanically vented at no less than six air changes per hour. Service corridors are required to have a means of exit. The maximum distance of travel from any point to a service corridor is not to exceed 75 feet. Dead ends cannot exceed 4 feet. There may not be less than two exits and no more than half can require travel into a fabrication area.

Service corridors must have doors that swing in the direction of the exit. The doors must have a minimum width of 5 feet or 33 inches wider than the widest cart or truck used in the corridor, whichever is greatest. As with required corridors, all service corridors are required to have an emergency alarm system in place.

Storage of hazardous production materials must be within approved or listed storage cabinets or gas cabinets within a workstation. Any HPM in quantities greater than those listed in Section 1804.2 of the International Fire Code must be stored in liquid storage rooms. The construction of HPM and gas rooms must be separated from other areas and by not less than a 2-hour fire barrier in an area that is 300 square feet (27.9 m²). Liquid storage rooms must be constructed in accordance with the following:

- Rooms in excess of 500 square feet (46.5 m²) must have at least one outer door approved for fire-department access.

- Rooms must be separated from other areas by fire barriers having a fire-resistance rating of not less than 1 hour for rooms up to 150 square feet (13.9 m²) in area and not less than 2 hours for larger rooms.

- Shelving, racks, and wainscoting in these areas must be of noncombustible construction or wood of not less than 1 inch (25 mm) normal thickness.

- Rooms used for the storage of Class I flammable liquids must not be in the basement.

!Codealert

Canopies used to support gaseous hydrogen systems require special consideration. See Section 406.5.2.1 for complete details.

Liquid-storage rooms and gas rooms must have a least one outside wall that is no less than 30 feet from lot lines adjacent to public ways. Explosion control is required. Where two exits are required, one must lead directly outside the building. Mechanical exhaust ventilation must be provided in liquid-storage rooms at the rate of no less than 1 cubic foot per minute (0.044 L/s/m^2) of floor area or six air changes per hour. HPM storage is required to have an approved emergency alarm system that is alarm-initiating, sounds a local alarm, and transmits a signal to the emergency control station.

Piping and tubing are required of systems that supply HPM liquids or gases having a health-hazard rating of 3 or 4 and must be welded throughout, except for connections to the system that are within a ventilated enclosure. Hazardous-production-materials supply piping in service must be exposed to view. Where HPM gases or liquids are carried in pressurized piping above 15 pounds per square inch in gauge (psig) (103.4 kPa), excess flow control must be provided. Where the piping comes from within a liquid-storage room, HPM room, or gas room, the excess flow control must be located within the room. The installation of HPM piping and tubing within the space must be in accordance with the following conditions:

• Automatic sprinklers must be installed within the space unless the space is less than 6 inches in the least dimension.

• Ventilation not less than six air changes per hour must be provided.

• The space must not be used to convey air from any other area.

• Where the piping or tubing is used to transport HPM liquids, a receptor must be installed below such piping or tubing.

• The receptor must be designed to collect any discharge or leakage and drain it to an approved location.

• The 1-hour enclosure must not be used as part of the receptor.

• HPM supply piping and tubing and nonmetallic waste lines must be separated from the corridor and from occupancies other than group H-5 by fire barriers that have a fire-resistance rating of not less than 1 hour.

• Where gypsum wallboard is used, joints on the piping side of the enclosure are not required to be taped, provided that the joints occur over framing members.

• Access openings into the enclosure must be protected by approved fire-protection-rated assemblies.

• Readily accessible manual or automatic remote-activation fail-safe emergency shutoff valves must be installed on piping and tubing other than waste lines at the following locations: (1) at branch locations into the fabrication area and (2) at entries into corridors. The exception to this is that any transverse crossing of the corridors by supply piping that is enclosed within a ferrous pipe or tube for the width of the corridor need not comply with items 1 through 5.

Gas Detection

Continuous gas-detection systems must be provided for HPM gases when the physiological warning threshold level of the gas is at a higher level than the accepted PEL for flammable gases in accordance with this section of the code. A continuous gas-detection system is required for fabrication areas, HPM rooms, gas cabinets, exhausted enclosures, and corridors but is not required for the occasional transverse crossing of the corridors by supply piping that is enclosed in a ferrous pipe or tube for the width of the corridor. The gas-detection system must be capable of monitoring the room and/or equipment in which the gas is located at or below the PEL or ceiling limit of the gas. For flammable gases, the monitoring detection threshold level shall be vapor concentrations in excess of 20 percent of the lower explosive limit (LEL). Monitoring for highly toxic gases must also comply with the requirements of the International Fire Code.

Gas-detection systems must initiate a local alarm and transmit a signal to the emergency-control station when short-term hazard is detected. The gas-detection system must automatically close the shutoff valve at the source on gas-supply piping and tubing related to the system being monitored for which gas or a short-term hazard is detected. Automatic closure of shutoff valves must comply with the following:

- Where the gas-detection sampling point initiating the gas-detection system alarm is within a gas cabinet or exhausted enclosure, the shutoff valve in the gas cabinet or exhausted enclosure for the specific gas detected must automatically close.
- Where the gas-detection sampling point initiating the gas-detection system alarm is within a room and compressed gas containers are not in gas cabinets or an exhausted enclosure, the shutoff valves on all gas lines for the specific gas detected must automatically close.
- Where the gas-detection sampling point initiating the gas-detection system alarm is within a piping distribution manifold enclosure, the shutoff valve supplying the manifold for the compressed-gas container of the specific gas detected must automatically close.

The only exception to this is where the gas-detection sampling point initiating the gas-detection system alarm is at the use location or within a gas-valve enclosure of a branch line downstream of a piping distribution manifold; the shutoff valve for the branch-line location in the distribution manifold enclosure must automatically close. The emergency-control station must receive signals from emergency equipment and alarm and detection systems. Such emergency equipment and alarm and detection systems must include but not be limited to the following where such equipment or systems are required to be provided either in this chapter or elsewhere in the code:

- Automatic fire-sprinkler system alarm and monitoring systems
- Manual fire-alarm systems
- Emergency alarm systems

- Continuous gas-detection systems
- Smoke-detection systems
- Automatic detection and alarm systems for phosphoric liquids and Class 3 water-reactive liquids as required in Section 1805.2.3.5 of the International Fire Code
- Exhaust-ventilation flow alarm devices for phosphoric liquids and Class 3 water-reactive liquids as required in Section 1805.2.3.5 of the International Fire Code

Emergency Power

An emergency power system must be provided for Group 5 occupancies. The required electrical systems are to include the following:

- HPM exhaust ventilation systems
- HPM gas-cabinet ventilation systems
- HPM exhausted enclosure ventilation systems
- HPM gas-room ventilation systems
- HPM gas detection systems
- Emergency alarm systems
- Manual fire-alarm systems
- Automatic-sprinkler-system monitoring and alarm systems
- Automatic alarm and detection systems for phosphoric liquids and Class 3 water-reactive liquids as required in Section 1805.2.3.5 of the International Fire Code
- Flow alarm switches for phosphoric liquids and Class 3 water-reactive liquids as required in Section 1805.2.3.5 of the International Fire Code
- Electrically operated systems as required elsewhere in the or in the International Fire Code applicable to the use, storage, or handling of HPM

Exhaust Ventilation

Exhaust ventilation systems are allowed to be designed to operate at no less than half the normal fan speed on the emergency power system where it is demonstrated that the level of exhaust will maintain a safe atmosphere. Automatic-sprinkler-system protection in exhaust ducts for HPM must be provided in exhaust ducts conveying gases, vapors, fumes, mists, or dusts generated by HPM. This applies to metallic and noncombustible nonmetallic exhaust ducts where the largest cross-sectional diameter is equal to or greater than 10 inches (254 c.m) and the ducts are within the building and convey flammable gases, vapors, or fumes, except for ducts listed or approved for applications without automatic fire-sprinkler-system protection or ducts not more than 12 feet (3,658 mm) in length installed below ceiling level.

Application of Flammable Liquids

This section applies to rooms where the spraying of flammable paints, varnishes, and lacquers or other flammable materials, mixtures, or compounds takes place. Spray rooms must be enclosed with fire barriers with no less than 1-hour fire-resistance rating. Floors must be waterproofed and drained in an approved manner. The interior surfaces of spray rooms must be smooth and constructed to permit the free passage of exhaust air. Spraying spaces must be ventilated to prevent the accumulation of flammable mist or vapors.

Drying Rooms

A drying room or kiln installed within a building must be constructed of approved noncombustible materials with overhead heating pipes that have a clearing of no less than 2 inches (5.1 cm) from combustible contents in the dryer. If the operating temperature of the dryer is 175 degrees F (79 degrees C) or more, metal enclosures must be insulated from adjacent combustible materials by no less than 12 inches (30.5 cm) of airspace, or the metal walls must be lined with 1/4-inch (0.64 cm) insulating mill board or other approved equivalent insulation. Drying rooms designed for high-hazard materials and processes, including special occupancies, must be protected by an approved automatic fire-extinguishing system complying with the provisions of Chapter 9.

Organic Coatings

Manufacturing of organic coatings must be done in a building that does not have pits or basements and must not occur in a building incidental to or connecting to buildings housing other operations. Mills that operate with close clearances and that process flammable and heat-sensitive materials, such as nitrocellulose, must be located in a detached building. All storage areas for flammable and combustible liquid tanks must be located at or above grade and must be separated from the processing area. Storage of nitrocellulose must be located on a detached pad or in a separate structure enclosed with no less than 2-hour fire barriers. Storage rooms for finished products that are flammable must be separated from the processing area by fire barriers with at least a 2-hour fire resistance.

Occupancies

Occupancies in Groups I-1, R-1, R-2, R-3 must comply with provisions of the code for these groups. Walls separating dwelling and sleeping units in the same building must comply with the provisions in Chapter 7. Floor/ceiling assemblies separating dwelling units in the same building must be constructed in accordance with provisions in Chapter 7.

Hydrogen Cutoff Rooms

Hydrogen cutoff rooms cannot be located below grade and must be classified with respect to occupancy in accordance with the provisions of Chapter 3 and separated from other areas of the building by not less than 1-hour fire barriers or as required by Chapter 5. Doors within fire-barrier walls, including doors to corridors, must be self-closing. Interior door openings must be electronically interlocked to prevent operation of the hydrogen system. When an exhaust system is used instead of the interlock system, the ventilation systems must work continuously and must be designed to work at negative pressure in relation to the surrounding area. The average velocity of ventilation at the face of the door opening with the door in the fully open position must not be less than 60 feet per minute (0.3048 mm/s with a minimum of 45 feet per minute at any point in the door opening. Windows must not be operable, and interior windows are not allowed. Cutoff rooms must be provided with mechanical ventilation and have an approved flammable-gas-detection system. Activation of the gas-detection system must result in initiation of an audible and visual alarm signal both inside and outside the cutoff room and activation of the mechanical ventilation system. Failure of the gas-detection system must shut down the mechanical ventilation system, cutting off hydrogen generation and sounding a trouble signal in an approved location. Explosion control must be provided in accordance with the International Fire Code. Mechanical standby ventilation and gas-detection systems must be connected to a standby power system in accordance with the provisions in Chapter 7.

!Codealert

The floor construction of the control area and the construction supporting the floor of the control area is allowed to be 1-hour fire-resistance rated in buildings of Type IIA, IIIA, and VA construction, provided that both of the following conditions exist:

1. The building is equipped throughout with an automatic sprinkler system in accordance with Section 903.3.1.1.
2. The building is three stories, or less, in height.

CHAPTER 5

GENERAL BUILDING HEIGHTS AND AREAS

In this chapter we will explore general building heights and areas. The chapter includes some definitions. The purpose of these definitions is to keep you updated on words that may take on different meanings in the 2006 International Building Code. I am going to introduce you to a few more groups and their classifications and the codes that these groups must follow.

DEFINITIONS

The following definitions are relevant to the building types discussed in this chapter:

• Area, Building: The area included within surrounding exterior walls (or exterior walls and fire walls) exclusive of vent shafts and courts. Areas of the building not provided with surrounding walls must be included in the building area if these areas are included within the horizontal projection of the roof or floor above.

• Basement: The portions of buildings that are partly or completely below grade plane. A basement must be considered as a story above grade plane where the finished surface of the floor above the basement is more than 6 feet (18.3 m) above grade plane or more than 12 feet (36.6 m) above the finished ground level at any point.

• Equipment Platform: An unoccupied, elevated platform used exclusively for mechanical systems or industrial process equipment, including the associated elevated walkways, stairs, and ladders necessary to access the platform (see Chapter 5 of the code).

• Grade Plane: A reference plane representing the average finished ground level adjoining the building at exterior walls. Where the finished ground level slopes away from the exterior walls, the reference plane must be established by the lowest points within the area between the building and the lot line or, where the lot line is more than 6 feet (1.8 m) from the building, between the building and a point 6 feet (1.8 m) from the building.

• Height, Building: The vertical distance from the grade plane to the average height of the highest roof surface.

• Height, Story: The vertical distance from top to top of two successive finished floor surfaces and, for the topmost story, from the top of the floor finish to the top of the ceiling joists or, where there is not a ceiling, to the top of the roof rafters.

• Mezzanine: An intermediate level or levels between the floor and ceiling of any story.

GENERAL HEIGHT AND AREA LIMITATIONS

The heights and areas of buildings of different construction types are governed by rules for the intended use of the building and cannot go above the limits in Table 4.4 (found in Chapter 4 of this book) except as provided in exceptions discussed below. Each part of a building included within the inside and outside walls and fire walls, where present, is allowed to count as a separate building. Buildings and structures that are designed to accommodate special industrial processes that require large areas and unusual heights to contain cranes or special machinery and equipment including, among others, the following, are exempt from the height and are limitations of Table 4.4:

• Rolling mills

• Structural metal fabrication shops and foundries

• The production and distribution of electric power

• The production and distribution of gas or steam power

There are situations when two or more buildings are on the same building lot. When this happens, they are regulated as separate buildings or considered as parts of one building if the height of each building and the total area of the buildings are within the limits of Table 4.4. The requirements for the total building are appropriate to each individual building. Buildings of Type I construction that are allowed to be of unlimited heights and areas are not subject to the special requirements that allow unlimited area buildings in Section 507 of the IBC or unlimited height in Sections 503.1.1 and 504.3 of the IBC.

Height

The height that is allowed by Table 4.4 can be increased in agreement with the provisions of this section of the code with the exception that the height of one-story aircraft hangars, aircraft paint hangers, and buildings used for the manufacturing of aircraft are not limited if the building is provided with an automatic fire-extinguishing system as dictated by Chapter 9 and entirely surrounded by public ways or yards no less in width than 1 1/2 times the height of the building.

Where a building is equipped throughout with an approved automatic sprinkler system in accordance with the provisions of Chapter 9, the value of Table 4.4 for maximum height is increased by 20 feet (6.1 m) and the maximum number of stories is increased by one story. For example, if Group R buildings are equipped throughout with an approved automatic sprinkler system in accordance with Chapter 9 and the value specified in Table 4.4, the maximum height is increased by 20 feet (6.1 m) and the maximum number of stories that can be increased is one. You cannot go over 60 feet (18.2 mm), or four stories, respectively. The following are exceptions:

• Fire areas with an occupancy in Group I-2 of Type IIB, III, IV, or V construction

• Fire areas with an occupancy in Group H-1, H-2, H-3, or H-5

• Fire-resistance-rating substitution in accordance with Table 4.2 (see note e)

Height includes roof structures. Towers, steeples, and other roof structures must be constructed of materials that are consistent with the required type of construction except where other construction is allowed by Chapter 15. These structures cannot be used for living space or storage. If roof structures, such as towers and steeples, are made of noncombustible materials, their height is unlimited, but they cannot extend more than 20 feet (6.1 m) above the allowable height if the structure is not made of noncombustible materials. Chapter 15 has additional information regarding these requirements.

Mezzanines

A mezzanine can be considered a portion of a story. It cannot, however, be counted toward the building area or the number of stories. The area of the mezzanine must be included in determining the fire area defined in Chapter 7. The clear floor height of the mezzanine cannot be less than 7 feet (2.1 m).

The total area of a mezzanine within a room is not allowed to be over one-third of the floor area of the room or the space that it is in. You cannot include the enclosed part of the room to determine the floor area where the mezzanine is located. The area of the mezzanine cannot be included in the floor area except for the following conditions:

• The total area of mezzanines in buildings and structures that are Type I or II for special industrial occupancies in accordance with this chapter of the code cannot be more than two-thirds of the area in the room.

• The total area of mezzanines in buildings and structures that are Type I or II cannot be more than one-half of the area of the room in buildings and structures that have an approved sprinkler system throughout. The sprinkler system has to be in

!Codealert

The aggregate area of mezzanines in buildings and structures of Type I or II construction shall not exceed one-half the area of the room in buildings and structures equipped throughout with an approved automatic sprinkler system in accordance with Section 903.3.1.1 and an approved emergency voice/alarm communication system in accordance with Section 907.2.12.2

accordance with Chapter 9, and an approved emergency voice/alarm communication system, also in accordance with Chapter 9, must be in place.

As with every building and structure that I've spoken about in this book, mezzanines have requirements for exits and exit routes. Each occupant of a mezzanine must have access to at least two exits where the common path of exit travel is over the limits of Chapter 10. If the exit from your mezzanine is a stairway, the maximum travel distance must include the distance traveled on the stairway as measured in the plane of the tread nosing.

Accessible means of exit have to be provided in accordance with Chapter 10, as well as a single means of exit. If your building or structure has a mezzanine, it has to be open, and no obstructions are allowed in the room where the mezzanine is located, except for walls that are not more than 42 inches (106.7 cm) high, columns, and posts.

There are five exceptions:

• Mezzanines or portions are not required to be open, provided that the occupant load is not more than 10 persons.

• Mezzanines or portions are not required to be open to the room if at least one of the exits provides direct access to an exit from the mezzanine level.

• Mezzanines are not required to be open to the room, provided that the total floor area of the enclosed space is not more than 10 percent of the area.

• In industrial facilities, mezzanines used for control equipment are allowed to be glazed on all sides.

• In other Groups H and I occupancies that are no more than two stories in height above grade plane and equipped with an automatic sprinkler, a mezzanine having two or more exits is not required to be open to the room in which the mezzanine is located.

!Codealert

In other than Groups H and I occupancies no more than two stories in height above grade plane and equipped throughout with an automatic sprinkler system in accordance with Section 903.3.1.1, a mezzanine having two or more means of egress shall not be required to be open to the room in which the mezzanine is located.

EQUIPMENT PLATFORMS

Equipment platforms in buildings cannot be considered as a portion of the floor below and must not contribute to either the building area or the number of stories, as regulated by this chapter of the code. You may also not use the area of the equipment platform to determine the fire area. Equipment platforms cannot be part of a mezzanine, and platforms, walkways, stairs, and ladders that provide access to an equipment platform cannot be used as an exit from the building. There are some area limitations that you must be aware of.

The total area of all equipment platforms within a room cannot be larger than two-thirds of the area of the room. If the equipment platform is located in the same room as a mezzanine, the combined total area of the equipment platform and mezzanine cannot be more than two-thirds of the room. If your mezzanine is in a building that is required to have an automatic sprinkler system, equipment platforms must be fully protected by these sprinklers above and below the platform where required by standards in Chapter 9. Equipment platforms must also have guards where required by Chapter 10.

Area Modifications

You know that Table 4.4 limits the areas of buildings and that areas are permitted to be increased with the proper automatic sprinkler system in place. Equation 5.1 must be applied when increasing these areas:

$$A_a = \{A_t + [A_1 \times I_f] + [A_1 \times I_s]\} \qquad \text{(Equation 5.1)}$$

where

s = Spacing of ribs or undulations

A_a = Allowable area per story in square feet

A_t = Tabular area per story in accordance with Table 4.4 in square feet

I_f = Area increase factor due to frontage as calculated in accordance with this chapter

I_s = Area increase factor due to sprinkler-system protection as calculated in accordance with this chapter

A single basement that is not a story above plane does need to be included in the total allowable area if it is not bigger than the area that is allowed for a building with no more than one story above grade plane.

Each building must connect or have access to a public way to receive an area increase for frontage. Where a building has more than 25 percent of its exterior on a public way or open space that has a minimum width of 20 feet (6.1 m), the frontage increase must be determined in accordance with equation 5.2.

$$I_f = [F/P - 0.25] \, W/30 \qquad \text{(Equation 5.2)}$$

where

I_f = Area increase due to frontage

F = Building perimeter that fronts on a public way or open space having 20 feet open minimum width

P = Perimeter of entire building in feet

W = Width of public way or open space (feet)

Width must be at least 20 feet. Where the value of width varies along the perimeter of the building, the calculation performed in accordance with equation 5.1 must be based on the weighted average of each part of the exterior wall and open space where the value of W is greater than or equal to 20 feet. When W is more than 30 feet (9.1 m), a value of 30 feet must be used in calculating the weighted average, regardless of the actual width of the open space. The exception is that the quantity of W divided by 30 must be a maximum of 2 when the building meets all requirements of this code chapter except for compliance with the 60-foot (18.2 m) public way or yard requirement, if applicable. These open spaces must be either on the same lot or dedicated for public use and must be accessed from a street or approved fire lane.

If a building is equipped with an approved automatic sprinkler system throughout the building and is in accordance with Chapter 9 provisions, the area is allowed to be increased by an additional 200 percent (Is = 2) for buildings with more than one story above grade plane and an additional 300 percent (Is = 3) for buildings with no more than one story above grade plane. These increases are permitted in addition to the height and story increases. The following list contains exceptions:

• The automatic-sprinkler-system increase does not apply to buildings with an occupancy in Group H-1.

• The automatic-sprinkler-system increase does not apply to the floor area of occupancy in Group H-2 or H-3. For mixed-use buildings containing these occupancies, the allowable area must be calculated, with the sprinkler increase applying only to the portions of the buildings not classified as Group H-2 or H-3.

• Fire-resistance-rating substitution in accordance with Table 4.2 (see note e)

The maximum area of a building with more than one story above grade plane must be determined by multiplying the allowable area of the first story (Aa), as determined in the above section, by the number of stories above grade plane as listed below:

• For buildings with three or more stories above grade plane, multiply by 3.

• For buildings with two stories above grade plane, multiply by 2.

• No story can be larger than the allowable area per story (Aa), as determined in the above section, for the occupancies on that story.

Unlimited-area buildings and the maximum area of buildings equipped throughout with an automatic sprinkler system determined by multiplying the allowable area per story (Aa) by the number of stories above grade plane are two exceptions to area determination.

In buildings with mixed occupancies, the allowable area must be based on the most restrictive provision for each occupancy. When the occupancies are treated as separate occupancies, the maximum total building area must be that of the ratios for each area on all floors and not more than 2 for two-story buildings and 3 for buildings three stories or higher.

Now that you have the appropriate equations and regulations for area and building modifications, you can increase the size of your buildings as needed. But don't forget one of many important rules: the building or structure must be equipped with automatic sprinkler systems that are in accordance with the provisions in Chapter 9.

Unlimited-Area Buildings

The area of buildings and occupancies and configurations of this section are not limited. The area of a one-story, non-sprinkler Group F-2 or S-2 building cannot be limited when the building is surrounded by and attached to public ways or yards not less than 60 feet (18.3 m) in width. The area of a one-story, Group B, F, M, or S building or a one-story Group A-4 building of other than Type V construction cannot be limited when the building is equipped with an automatic sprinkler system throughout, provided that it is in accordance with hapter 9 and is surrounded and attached by public ways or yards not less than 60 feet (18.3 m) in width. Exceptions include:

• Buildings and structures of Type I and II construction for rack storage facilities that do not have access by the public cannot be limited in height, provided that these buildings conform to the requirements of Chapters 5 and 9 and NFPA 230.

• The automatic sprinkler system is not required in areas occupied for indoor participant sports, such as tennis, skating, swimming, and equestrian activities in occupancies in Group-A if exit doors to the outside are provided and the building is equipped with a sprinkler system in accordance with Chapter 9.

• Group A-1 and A-2 occupancies other than Type V construction are permitted, provided that all assembly occupancies are separate from other spaces as required with no reduction in the fire-resistance rating, each Group A occupancy is not larger than the maximum allowance, and all required exits lead directly outdoors.

If a two-story building is classed as Group B, F, M, or S and is equipped with an automatic sprinkler system that is in accordance with Chapter 9, the area is not limited. It must, however, be surrounded and adjoined by public ways or yards no les than 60 feet (18 m, 288 mm) in width. The permanent open space of 60 feet required in this chapter will be permitted to be reduced to no less than 40 feet (12.1 m) if the following requirements are met:

• The reduced open space is not allowed for more than 75 percent of the perimeter of the building.

• The exterior wall facing the reduced open space must have a minimum fire-resistance rating of 3 hours.

• Openings in the exterior wall facing the reduced open space must have opening protectives with a minimum fire protection rating of 3 hours.

The area of a one-story, Group A-3 building used as one of the following is permitted to be unlimited:

• A place of religious worship
• Community hall
• Dance hall
• Exhibition hall
• Gymnasium
• Lecture hall
• Indoor swimming pool
• Tennis court

These group uses must be of Type II construction and are not limited when the building does not have a stage other than a platform if the building is equipped throughout with an automatic sprinkler system, the assembly floor is located at or within 21 inches (53.3 cm) of street or grade level, all exits are provided with ramps complying with Chapter 10, and the building is not surrounded and adjoined by public ways or yards no less than 60 feet (18.2 m) in width.

Group H occupancies including H-2, H-3, and H-4 occupancies arepermitted in unlimited-area buildings containing Group F and S occupancies. The total floor area of the Group H occupancies that are located outside the unlimited-area building cannot be more than 10 percent of the area of the building or the area limitations for Group H occupancies specified in Table 4.4 based upon the percentage of the perimeter of each Group H fire area that looks onto on a street or other unoccupied space. The total floor area of Group H occupancies not located at the outside of the building cannot be more than 25 percent of the area limitations for the Group H occupancies as specified in Table 4.4. For two-story unlimited-area buildings, the Group H fire areas cannot be located above the first story unless permitted by the allowable height in stories and feet as described in Table 4.4 based on the type of construction of the unlimited-area building.

The area of a one-story, Group H-2 aircraft paint hangar is not limited when it complies with the provisions of Chapter 4 and is totally surrounded by public ways or yards no less in width than 1 1/2 times the height of the building.

Group E buildings of Type II, IIIA, or IV construction are not limited when the following criteria are met:

• Each classroom must have no less than two means of exits, with one of those exits being a direct exit to the outside of the building in compliance with Chapter 10.
• The building is equipped throughout with an automatic sprinkler system in accordance with Chapter 9.
• The building is surrounded by and joined by public ways or yards no less than 60 feet (18.2 m) in width.

Motion-picture theaters of one story and of Type II construction are not limited when the building is provided with an automatic sprinkler system throughout in accordance with Chapter 9 and is surrounded and connected by public ways or yards no less than 60 feet (18 m, 288 mm) in width.

Covered mall buildings and anchor stores having no more than three stories in height that comply with Chapter 4 are not limited.

Mixed-Use Occupancy

Buildings or parts of buildings that contain two or more occupancies or uses are classified as mixed-use. Incidental uses must comply with this section, the only exception being areas that serve as dwelling or living spaces. An incidental-use area must be classified in accordance with the occupancy of that portion of the building in which it is located, or the building must be classified as a mixed occupancy and must comply with the provisions of this section. In Table 5.1 where a fire-resistance-rated separation is required, the incidental-use area must be separated from the rest of the building by a fire barrier or a horizontal assembly constructed in accordance with Chapter 7.

Where Table 5.1 allows an automatic fire-extinguishing system without a fire barrier, the incidental-use area must be separated from the rest of the building by construction that is capable of preventing smoke from passing through the building. The partitions must extend from the floor to the underside of the fire-resistance-rated floor/ceiling assembly or fire-resistance-rated roof/ceiling assembly above or to the bottom of the floor or roof sheathing or subdeck above. Doors must

TABLE 5.1 Incidental use areas.

ROOM OR AREA	SEPARATION AND/OR PROTECTION
Furnace room where any piece of equipment is over 400,000 Btu per hour input	1 hour or provide automatic fire-extinguishing system
Rooms with boilers where the largest piece of equipment is over 15 psi and 10 horsepower	1 hour or provide automatic fire-extinguishing system
Refrigerant machinery rooms	1 hour or provide automatic sprinkler system
Parking garage (Section 406.2)	2 hours; or 1 hour and provide automatic fire-extinguishing system
Hydrogen cut-off rooms, not classified as Group H	1 hour in Group B, F, M, S and U occupancies. 2-hour in Group A, E, I and R occupancies.
Incinerator rooms	2 hours and automatic sprinkler system
Paint shops, not classified as Group H, located in occupancies other than Group F	2 hours; or 1 hour and provide automatic fire-extinguishing system
Laboratories and vocational shops, not classified as Group H, located in Group E or I-2 occupancies	1 hour or provide automatic fire-extinguishing system
Laundry rooms over 100 square feet	1 hour or provide automatic fire-extinguishing system
Storage rooms over 100 square feet	1 hour or provide automatic fire-extinguishing system
Group I-3 cells equipped with padded surfaces	1 hour
Group I-2 waste and linen collection rooms	1 hour
Waste and linen collection rooms with over 100 square feet	1 hour or provide automatic fire-extinguishing system
Stationary lead-acid battery systems having a liquid capacity of more than 100 gallons used for facility standby power, emergency power or uninterrupted power supplies	1 hour in Group B, F, M, S and U occupancies. 2-hour in Group A, E, I and R occupancies

For SI: 1 square foot = 0.0929 m^2, 1 pound per square inch = 6.9 kPa, 1 British thermal unit per hour = 0.293 watts, 1 horsepower = 746 watts, 1 gallon = 3.785 L.

!Codealert

Mixed-use occupancy regulations in Section 508 have many new elements compared to previous code requirements.

be self-closing or automatic-closing when smoke is detected. Doors must not have any air-transfer openings and cannot be undercut in excess of the clearance that is permitted in accordance with NFPA 80.

Where an automatic fire-extinguishing system or automatic sprinkler system is provided in accordance with Table 5.1, only the incidental-use areas need to be equipped.

Mixed occupancies must comply with the provisions of this section either at an individual level or a combination of the following sections. However, occupancies that are separate and in accordance with the special provisions section and occupancies that are required to comply with Table 4.11 and are areas of Group H-1, H-2, or H-3 occupancies that must be located in separate and detached buildings are an exception.

Accessory Occupancies

Accessory occupancies are occupancies secondary to the main occupancy of the building or portion of the building. Total accessory occupancies will not occupy more than 10 percent of the area of the story that they are located in and cannot have more than the tabular values in Table 4.4 without height and area increases in accordance with rules for exceptions. There are three exceptions:

• Accessory assembly areas having a floor area less than 750 square feet (69.7 m²) are not considered separate occupancy.

• Assembly areas that are accessory to Group E occupancies are not considered separate occupancies except when applying the assembly occupancy requirements of Chapter 11.

• Accessory religious-educational rooms and religious auditoriums with occupant loads of less than 100 are not considered separate occupancies.

Occupancy classifications require that accessory occupancies must be individually classified in accordance with Chapter 3. Code requirements must apply to each section of the building based on the occupancy classification of that accessory space, except that the most restrictive or strictest applicable provisions of

!Codealert

Section 509.8 refers to Group B or M with Group S-2 open parking garage above has many new code changes.

Chapter 4 and Chapter 9 must apply to the entire building or portions of the building.

The allowable area and height of the building must be based on the allowable area and height for the main occupancy in accordance with this chapter of the code. The height of any accessory must not be more than the tabular values in Table 4.4 without height and area increases in accordance with the chapter for accessory occupancies.

No separation is needed between accessory occupancies and the main occupancy except for Group H-2, H-3, H-4, and H-5 occupancies, which must be separated from all other occupancies i.

Non-separated occupancies must comply with the provisions of this section and must be individually classified in accordance with Chapter 3. Code requirements will apply to each section of the building based on the occupancy classification of that space except that the strictest provisions of Chapter 4 and Chapter 9 must apply to the entire building or portion of the building used.

The allowable area and height of the building or portion that is used must also be based on the strictest allowances for the occupancy groups under consideration for the type of construction of the building. No separation is required between occupancies except Groups H-2, H-3, H-4, and H-5 occupancies, which must be separated from all other occupancies. Separate occupancies must be individually classified in accordance with Chapter 3. Each fire area must comply with the occupancy classification of that portion of the building. In each story, the building area must be the sum of the ratios of the actual floor area of each occupancy divided by the allowable area of each occupancy but cannot be more than one.

Each occupancy must comply with height limitations based on the type of construction of the building. The height, in both feet and stories, of each fire area is measured from grade plane; this measurement has to include the height, in both feet and stories, of intervening fire areas, except for the special-provisions section of this chapter of the code. Separation requires that individual occupancies must be separated from nearby occupancies in accordance with Table 5.2, and separations that are required must have fire barriers or horizontal assemblies constructed in accordance with Chapter 7, or both, so that complete separation of occupancy occurs.

Special Provisions

This section contains special provisions and special conditions that are exempt from or modifications of the specific requirements of this chapter regarding the allowable heights and areas of buildings based on the occupancy classification and type of construction, provided that the special condition complies with the provisions specified in this section for each condition and other requirements that are required.

Let's start with Group S-2, which, if you recall, is classified as enclosed or open parking garages and includes Group A, B, M, R, and S. A basement and/or the first story above grade plane of a building must be considered as a separate and distinct building for the purpose of determining area limitations, continuity of fire walls, limitation of number of stories, and type of construction when the following are met.

The basement and/or the first story above grade plane is of Type IA construction and is separated from the building above with a horizontal assembly having a minimum 3-hour fire-resistance rating.

Shaft, stairway, ramp, or escalator enclosures through the horizontal assembly must have no less than a 2-hour fire-resistance rating with opening protectives in accordance with Table 5.3.

Exceptions to this are as follows:

• Where the enclosure walls below the horizontal assembly have less than a 3-hour fire-resistance rating in accordance with Table 5.3, the enclosure walls extending above the horizontal assembly must have a 1-hour fire-resistance rating, provided the building above the horizontal assembly is not required to be of Type 1 construction, the enclosure connects fewer than four stories, and the enclosure opening protectives above the horizontal assembly have a minimum 1-hour fire-protection rating.

• The building above the horizontal assembly must have multiple Group A uses, each with an occupant load of less than 300 or Group B, M, R, or S uses.

• The building below the horizontal assembly is a Group S-2 enclosed or open parking garage, used for the parking and storage of private motor vehicles.

Exceptions include the following:

• Entry lobbies, mechanical rooms, and similar uses incidental to the operation of the building will be permitted.

• Multiple Group A uses, each with an occupant load of less than 300, or Group B or M uses will be permitted, in addition to those uses incidental to the operation of the building (including storage areas), provided that the entire structure below the horizontal assembly is protected throughout by an approved automatic sprinkler system.

• The maximum building height in feet will not be more than the limits allowed in

TABLE 5.2 Required separation of occupancies (hours).

Occupancy	A^e,E		I		R^d		F-2,S-2^c,d,U^d		B^b,F-1,M^b,S-1		H-1		H-2		H-3,H-4,H-5	
	S	NS	S	NS	S	NS	S	NS	S	NS	S	NS	S	NS	S	NS
A^e,E^e	N	N	1	2	1	2	N	1	1	2	NP	NP	3	4	2	3^a
I	—	—	N	N	1	NP	1	2	1	2	NP	NP	3	NP	2	NP
R^d	—	—	—	—	N	N	1	2	1	2	NP	NP	3	NP	2	NP
F-2,S-2^c,d,U^d	—	—	—	—	—	—	N	N	1	2	NP	NP	3	4	2	3^a
B^b,F-1,M^b,S-1	—	—	—	—	—	—	—	—	N	N	NP	NP	2	3	1	2^a
H-1	—	—	—	—	—	—	—	—	—	—	N	NP	NP	NP	NP	NP
H-2	—	—	—	—	—	—	—	—	—	—	—	—	N	NP	1	NP
H-3,H-4,H-5	—	—	—	—	—	—	—	—	—	—	—	—	—	—	N	NP

For SI: 1 square foot = 0.0929 m².

S   = Buildings equipped throughout with an automatic sprinkler system installed in accordance with Section 903.3.1.1.
NS = Buildings not equipped throughout with an automatic sprinkler system installed in accordance with Section 903.3.1.1.
N   = No separation requirement.
NP = Not permitted.

a. For Group H-5 occupancies, see Section 903.2.4.2.
b. Occupancy separation need not be provided for storage areas within Group B and M if the:
    1. Area is less than 10 percent of the floor area;
    2. Area is equipped with an automatic fire-extinguishing system and is less than 3,000 square feet; or
    3. Area is less than 1,000 square feet.
c. Areas used only for private or pleasure vehicles shall be allowed to reduce separation by 1 hour.
d. See Section 406.1.4.
e. Commercial kitchens need not be separated from the restaurant seating areas that they serve.

this chapter for the building having the smaller allowable height as measured form the grade plane.

- A Group S-2 enclosed parking garage located in the basement or first story below a Group S-2 open parking garage will be classified as a separate and distinct building for the purpose of determining the type of construction when the following conditions are met:

  - The allowable area of the structure must be such that the sum of the ratios of the actual area divided by the allowable area for each separate occupancy will not be above 1.0.

  - The Group S-2 enclosed parking garage is of Type I or II construction and is at least equal to the fire-resistance requirements of the Group S-2 open parking garage.

  - The height and the number of floors above the basement must be limited as specified in Table 4.3.

  - The floor assembly separating the Group S-2 enclosed parking garage and Group S-2 open parking garage must be protected as required for the floor assembly of the Group S-2 enclosed parking garage. Openings between the Group S-2 enclosed parking garage and Group S-2 open parking garage, except exit openings, are not required to be protected.

  - The Group S-2 enclosed parking garage is used exclusively for the parking or storage of private motor vehicles but may contain an office, waiting room, and bathroom having a total area of not more than 1,000 square feet (93 m²) and the mechanical equipment rooms incidental to the operation of the building.

Where a maximum one-story above-grade-plane Group S-2 parking garage, enclosed or open, a combination of Type I construction or open Type IV construction, with grade entrance, is provided under a building of Group R, the number of stories to be used in determining the minimum type of construction must be measured from the floor above such a parking area. The floor assembly between the parking garage and the Group R building must comply with the type of construction required for the parking garage and must also provide a fire-resistance rating no less than the mixed-occupancy requirement in this chapter of the code.

The height limitation for buildings of Type IIIA construction in Group R-2 must be increased to six stories and 75 feet (22.9 m) where the first-floor construction above the basement has a fire-resistance rating of not less than 3 hours and the floor area is subdivided by 2-hour fire-resistance-rated fire walls into areas of no more than 3,000 square feet (279 m²). The height limitations for building of Type IIA differ in that construction in Group R-2 must be increased to nine stories and 100 feet (30/5 m) where the building is separated by no less than 50 feet (15.2 m) from any other building on the lot and from lot lines. The exits must be segregated in an area enclosed by a 2-hour fire-resistance-rated fire wall and the first-floor construction has a fire-resistance rating of no less than 1 1/2 hours.

Open parking garages that are beneath Groups A, I, B, M, and R buildings cannot exceed the height and area limitations permitted in Chapter 4. The height and area of the portion of the building above the open parking garage cannot exceed the limitations in this chapter for the upper occupancy. The height, in both feet and stories, of the portion of the building above the open parking garage must be measured from grade plane and must include both the open parking garage and the portion of the building above the parking garage.

Fire barriers or horizontal assemblies constructed in accordance with Chapter 7 between the parking occupancy and the upper occupancy must correspond to the required fire-resistance rating prescribed in Table 5.2. The type of construction must apply to each occupancy individually, except that structural members, including main bracing within the open parking structure, which is necessary to support the occupancy on top, must be protected with the more restrictive fire-resistance-rated assemblies of the groups involved as shown in Table 4.2. Exits for the top occupancy must conform to Chapter 10 and must be separated from the parking occupancy by fire barriers having at least a 2-hour fire-resistance rating or horizontal assemblies having at least a 2-hour fire-resistance rating with self-closing doors in compliance with Chapter 7. Exits from the open parking garage must comply with Chapter 4.

Group B or M uses located in the basement or first story below a Group S-2 open parking garage must be classified as separate and distinct buildings for the purpose of determining the type of construction when all the following conditions are met:

- The basement or first story must be Type I or II construction but no less than the type of construction required for the open parking garage above. The height and area of the basement or first story cannot exceed the limitations for the Group B or M uses.

- The height and area of the open parking garage cannot exceed the limitations permitted in Chapter 4. The height, in both feet and stories, of the open parking garage must be measured from grade plane and include both the open parking garage and the basement or first story.

- Fire separation assemblies between the open parking garage and the basement or first story must correspond to the required fire-resistance rating in Table 5.2.

- Exits in the open parking garage must go directly to a street or public way and must be separated from the basement or first story by not less than 2-hour fire barriers or 2-hour horizontal assemblies constructed in accordance with Chapter 7, or both, with opening protectives in accordance with Table 5.3.

Now we are ready to move into Chapter 6 to expand our knowledge of code requirements.

> **!Code**alert
>
> In buildings of Type I construction exceeding two stories in height, fire-retardant-treated wood is not permitted in roof construction when the vertical distance from the upper floor to the roof is less than 20 feet.

ceilings loads only. Columns must be continuous or superimposed and connected in an approved manner. Type V construction is that type of construction that contains structural elements, exterior and interior walls. You will be reading more about this type of construction in the section regarding roof decks on Page 6.6.

FLOOR FRAMING

Floor framing, wood beams, and girders must be sawed or glue-laminated timber and cannot be less than 6 inches (16.2 cm), commercial-sized, in width and less than 10 inches (25.4 cm), commercial-sized, in depth. Framed sawed or glue-laminated timber arches, which spring from the floor line and support floor loads, cannot be less than 8 inches (20.3 cm), commercial-sized, in any dimension. Framed lumber trusses supporting floor loads cannot be less than 8 inches (20.3 cm), commercial-sized, in any dimension.

ROOF FRAMING

Roofs can have wood-frame or glue-laminated arches for construction. If they spring from the floor line or from grade and do not support floor loads, they must have members not less than 6 inches (15.2 cm), commercial-sized, in width, not less than 8 inches (20.3 cm), commercial-sized, in depth for the lower half of the height, and not less than 6 inches (15.2 cm), commercial-sized, in depth for the upper half. Framed or glued-laminated arches that spring from the top of walls or wall abutments, framed timber trusses, and other roof framing, which do not support floor loads, cannot have members less than 4 inches (10.2 cm), commercial-sized, in width and not less than 6 inches (15.2 cm), commercial-sized, in depth. Spaced members may be composed of two or more pieces not less than 3 inches (7.6 cm), commercial-sized, in thickness secured to the underside of the members.

!Codealert

Sprayed fire-resistant materials and intumescent and mastic fire-resistant coatings, determined on the basis of fire-resistance tests in accordance with Section 703.2 and installed in accordance with Section
1704.10 and 1704.11, respectively.

Splice plates cannot be less than 3 inches (7.6 cm), commercial-sized, in thickness
secured to the underside of the members. Where protected by approved automatic
sprinklers under the roof deck, framing members cannot be less than 3 inches (7.6
cm), commercial-sized, in width.

FLOORS

Floors must be without concealed spaces. Wood floors must be of sawed or glued-
laminated planks, splined or tongue-and-groove, not less than 3 inches (7.6 cm),
commercial-sized, in thickness covered with 1-inch (2.5 cm), commercial-sized,
dimensioned tongue-and-groove flooring, laid crosswise or diagonally, or 0.5-
inch (1.3 cm) particleboard or planks not less than 4 inches (10.2 cm), commercial-sized, in width set on edge close together, well spiked, and covered with 1-
inch (2.5 cm), commercial-sized, dimension flooring, 15/32-inch (1.2 cm) wood
structural paneling, or 0.5-inch (1.3 cm) particleboard. It is very important that
you lay the lumber so that no continuous line of joints will occur except at points
of support. Floors cannot extend closer than 0.5 inch (1.3 cm) to walls, and the
space must be covered by a molding fastened to the wall and so arranged that it
will not cover the swelling or shrinkage movements of the floor. Corbeling or
building out of masonry walls under the floor can be used instead of traditional
molding.

ROOF DECKS

Roofs must have no hidden spaces; any wood roof decks must be sawed, glued-
laminated, splined, or tongue-and-groove plank no less than 2 inches (5.1 cm)
nominal thickness, 1 1/8-inch-thick (3.2 cm) wood structural panel (exterior glue)
or of planks not less than 3 inches (7.6 cm) nominal width, set on edge close together and laid as required for floors. You can use other types of decking if you

provide the equivalent fire resistance and structural properties. Partitions need to be of solid wood construction, formed of at least two layers of matched boards or laminated construction 4 inches (10.2 cm) thick or of 1-hour fire-resistance-rated construction. The external structural members need to be wood columns and arches with a horizontal separation of 20 feet (61 m). The wood columns and arches need to be made of heavy timber that has been treated for outside use. All of the structural elements that I have talked about in this chapter such as exterior and interior walls are considered to be of Type V construction and are allowed by this code.

COMBUSTIBLE MATERIALS

Type I and Type II buildings and structures allow the use of combustible materials. But always remember to follow the applications in this section of the code.

Fire-retardant-treated wood is allowed to be used in the following instances:

• Nonbearing partitions where the required fire-resistant rating is 2 hours or less

• Nonbearing exterior walls where no fire rating is required

• Roof construction that includes girders, trusses, framing, and decking, except for Type I buildings more than two stories in height; fire-retardant-treated wood is not allowed in roof construction when the vertical distance from the upper floor to the roof is less than 20 feet (6.1 m)

• Thermal and acoustical insulation, other than foam plastics, having a flame spread index of not more than 25, except insulation placed between two layers of noncombustible materials without an intervening airspace, which can only have a flame spread index of no more than 100, and insulation between a finished floor and solid decking without intervening airspace, which can have a flame spread index of no more than 200.

• Foam plastics used in accordance with Chapter 26 of the code

• Roof coverings that have an A, B, or C classification

• Interior floor finishes, interior finish, trim, and millwork, such as doors, door frames, window sashes, and window frames

• Where not installed more than 15 feet (4.6 m) above grade, show windows, nailing or furring strips, and wooden bulkheads below show windows, including their frames, aprons and show cases

• Finished flooring applied directly to the floor slab or to wood sleepers that are fire-blocked in accordance with Chapter 7 provisions

• Partitions up to 6 feet (1.8 m) in height that divide portions of stores, offices, or similar places occupied by one tenant only and that do not produce a corridor serving an occupant load of 30 or more

• Stages and platforms constructed in accordance with Chapter 4 provisions

- Combustible outside wall coverings, balconies, and similar projections and bay or oriel windows built in accordance with Chapter 14 provisions
- Blocking for handrails, millwork, cabinets, and window and door frames
- Light-transmitting plastics as specified in Chapter 26
- Mastics and caulking materials applied to provide flexible seals between components of outside wall construction
- Outside plastic veneer installed in accordance with Chapter 26 provisions
- Nailing or furring strips as allowed by Chapter 8
- Heavy timber as permitted by Chapters 6 and 14
- Aggregates, component materials, and admixtures as permitted by Chapter 7
- Sprayed fire-resistant materials and fire-resistant coatings determined on the basis of fire-resistance tests in accordance with Chapter 7 and installed in accordance with Chapter 17
- Materials used to protect penetrations in fire-resistance-rated assemblies in accordance with Chapter 17
- Materials allowed in the concealed spaces of buildings of Type I and II construction in accordance with Chapter 7
- Materials exposed within plenums complying with section 602 of the International Mechanical Code

The International Mechanical Code allows the use of nonmetallic ducts when installed in accordance with the code. The International Mechanical Code and the International Plumbing Code permit the use of combustible piping materials when installed in compliance with their provisions.

The ICC Electrical Code allows the use of electrical wiring with combustible insulation, tubing, raceways, and related components when installed improperly.

CHAPTER 7
FIRE-RESISTANCE-RATED CONSTRUCTION

The definition-alert boxes in this chapter contain words that may have a slightly different meaning in building construction than what you are used to. They are located throughout this chapter; look at them, read them, and know them. They are here to help you. Let's start with one now:

> # !Definitionalert
>
> **Annular Space:** The opening around a penetration item, such as a wall, floor or ceiling.

FIRE-RESISTANCE RATING AND FIRE TESTS

The fire-resistant materials that I describe in this chapter conform to the requirements of the chapter in your code book. The fire-resistance rating of building elements is determined in accordance with test procedures required by ASTM E 119 or in accordance with your local code. If you have materials, systems, or devices that have not been tested as part of a fire-resistance-rated assembly but are incorporated into the assembly, you must show sufficient data to the building official that proves that the required fire-resistance-rating is present. All the materials and methods that produce joints and penetration in fire-resistance-rated buildings must achieve the required fire-resistance rating except for one condition. When you are determining the fire-resistance rating of outside walls, compliance with the ASTM

E 119 data for unexposed surface temperature rise and ignition of cotton waste due to flame or gas is only required for the time period that corresponds to the required fire-resistance rating of an outside nonbearing wall that has the same fire separation distance and that is in a building of the same group.

If the inside walls of your building or structure are not proportioned or even, they have to be tested with both faces exposed to the furnace and the assigned fire-resistance rating has to be the shortest duration obtained from the two tests that you conduct in agreement with ASTM E 119. If the evidence given to the building official shows that the wall was tested with the least fire-resistant side exposed to the furnace, approval is up to the code officer's discretion. If the building official does give you a green light, then you are in luck, because the walls will not need to be tested from the opposite side.

Combustible components such as aggregates are permitted in gypsum and cement-concrete mixtures approved for fire-resistance-rated construction. Any approved element or admixture (the product of two substances mixed together) is allowed in assemblies if the tested assembly meets the fire-resistance test requirements. These assemblies that are tested under ASTM E 119 cannot be considered resistant to thermal effects unless the evidence that you present is in accordance with ASTM E 119 and approved by the building official. But do not forget that any and all restrained construction must be put on your plans. The following methods are alternatives for determining fire-resistance-rating, but they must conform to ASTM E 119 and be approved by your building official:

• Fire-resistance designs documented in approved sources

• Prescriptive designs of fire-resistance-rated building elements as approved in this chapter of the code

• Calculations in accordance with this chapter of the code

• Engineering analysis based on comparing building-element designs that have fire-resistance ratings determined by ASTM E 119

• Alternative protection methods as discussed in Chapter 1 of the code

!Definitionalert

Fire-Resistance Rating: The period of time a building element, component, or assembly maintains the ability to confine a fire, continues to perform a given structural function, or both, as determined by the tests or methods described in this chapter of the code.

!Codealert

A fire-resistance-rated wall assembly of materials designed to restrict the spread of fire in which continuity is maintained is a fire barrier.

Any test that you use in this chapter has to be done with acceptable materials that were specified in Chapter 6. When I use the word "noncombustible," understand that it does not apply to the flame-spread characteristics of inside finish or trim materials. If the material that you are using is subject to an increase in combustibility or flame spread beyond the limits that are allowed, it must not be considered noncombustible. There are two types of materials, elementary and composite materials, that are noncombustible; they do not conform to ASTM E 119. Elementary materials that are required to be noncombustible must be tested in accordance with ASTM E 136, and composite materials that have a structural base of noncombustible materials with a surface thickness of no more than 0.125 inch (3.18 mm) and a flame-spread index of no more than 50 when tested in accordance with ASTM E 84 will be acceptable as noncombustible materials.

EXTERIOR WALLS

Exterior walls are walls, bearing or nonbearing, used as enclosing walls. They cannot be fire walls and must have a slope of 60 degrees (1.01 rad) or greater with the horizontal plane. This section of the code includes moldings, eave overhangs, outside balconies, and similar projections that extend beyond the floor area. Outside balconies and stairways that have exits must also comply with Chapter 10. There are two methods to determine the allowable distance that your walls may extend. The method that allows the least amount of distance is the one to use.

!Definitionalert

Horizontal Assembly: A fire-resistance-rated floor or roof assembly of materials designed to restrict the spread of fire from the top of the foundation or floor to the underside of the roof sheathing.

These methods are: (1) a point one-third the distance to the lot line from an assumed vertical plane located where protected openings are required in accordance with this chapter, (2) more than 12 inches (30.5 cm) into areas where openings are prohibited. If the building or structural wall projections are of Type I or Type II construction, they have to be made of materials that are noncombustible or of combustible types allowed by Chapter 14.

Projections from walls that are Type III, IV, or V construction can be of any approved material. With the exception of Type V construction, which is allowed for R-3 occupancies, all combustible projections that are located where openings are not allowed or where protection of openings is required must have at least a 1-hour fire-resistance-rating.

SHARED LOTS

Let me touch upon the requirements of buildings that are on the same lot. It can get tricky when there are two buildings on the same lot. To make things a little less complicated, you must imagine an imaginary line separating the two buildings so that the outside walls and protective openings will be separated, at least in your mind. You may be questioning whether these buildings could possibly have two separate occupancies. This should be of no concern as long as the area allows for the more restrictive occupancy and the outside walls are constructed of materials that are permitted for that type of occupancy.

Each outside wall has to be fire-resistance-rated in accordance with Tables 6.1 and 6.2. Each wall with a distance more than 5 feet (1.5 m), has to be rated for exposure to fire from the back or inside. If the distance is less than 5 feet, the rating must be for exposure from both sides, and the walls must have sufficient stability so that they will remain standing for the required time as indicated by the fire-resistance rating.

If the building has protected openings that are not limited in area by this chapter, then the limitation on the rise of temperature on the unexposed surface of outside walls that ASTM E 119 requires will not apply. If the protected openings are limited in area as required by ASTM E 119, the limitation will not apply except under this condition: a correction must be made for radiation from the unexposed outside-wall surface using the following equation (see Figure 7.1):

where:

A_e = Equivalent area of protected openings

A = Actual area of protected openings

A_f = Area of outside wall surface in the level or story of the building that is under consideration regarding openings on which the temperature limitations of ASTM E 119 are exceeded

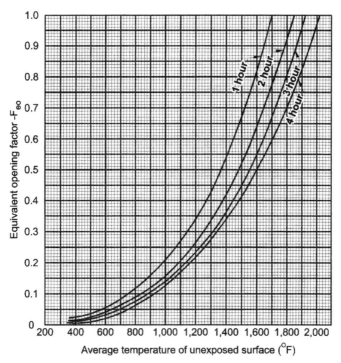

FIGURE 7.1 Equivalent opening factor.

Feo = The equivalent opening factor based on the average temperature of the unexposed wall surface and the fire-resistance rating of the wall

I know what you are thinking—this is why I've given you a graph to consult. But I am not done yet. I need to go into the allowable area of openings, which includes a formula and a table for better understanding.

!Definitionalert

Fire Wall: A fire-resistance-rated wall having protected openings, restricting the spread of fire, and extending continuously from the foundation to or through the roof, with sufficient structural stability under fire conditions to allow collapse of construction on either side without collapse of the wall.

The maximum area of any unprotected or protected openings that are allowed in outside walls in any story or level of the building cannot go over the values in Table 7.1. The total area of these openings has to comply with this formula:

where:

A = Actual or equivalent area of protected openings Ae

a = Allowable area of protected openings

A_u = Actual area of unprotected openings

a_u = Allowable area of unprotected openings

The graph and the table will make it much easier for you to understand the equations when you see how they are used in conjunction for outside wall allowances. As with other elements that require an automatic sprinkler system, unprotected openings are no different, unless, as according to Chapter 9, your occupancy is one of Group H-1, H-2, or H-3. On the first story or level in occupancies other than Group H, you may have an unlimited amount of unprotected openings in the outside walls that are above grade facing a street and that have a fire-separation distance more than 15 feet (4.6 m) and face an unoccupied space. But there are rules for this allowance too. This space must be on the same lot or be dedicated for public use. It has a size restriction of less than 30 feet (9.1 m) in width, and it must have access from a street with a posted fire lane as required by the International Fire Code.

OUTSIDE WALLS

Outside walls, as I have discussed, have protected and unprotected openings. These openings are required to be vertically separated so that they are protected against fire spread on the outside of the buildings if they are within 5 feet (1.5 m) of each other horizontally and the opening in the lower story is not a pro-

> **!Code**alert
>
> Fire-separation distance is the distance measured from the building face to one of the following:
> - The closet interior lot line
> - To the centerline of a street, an alley, or public way
> - To an imaginary line between two buildings on the property

TABLE 7.1 Maximum area of exterior wall openings[a].

CLASSIFICATION OF OPENING	FIRE SEPARATION DISTANCE (feet)							
	0 to 3[i,j]	Greater than 3 to 5[c,g]	Greater than 5 to 10[c,e,g,h]	Greater than 10 to 15[d,e,g]	Greater than 15 to 20[d,i,g]	Greater than 20 to 25[d,g]	Greater than 25 to 30[d,g]	Greater than 30
Unprotected	Not Permitted	Not Permitted[c]	10%[i]	15%[i]	25%[i]	45%[i]	70%[i]	No Limit[b]
Protected	Not Permitted	15%	25%	45%	75%	No Limit[b]	No Limit[b]	No Limit[b]

For SI: 1 foot = 304.8 mm.

a. Values given are percentage of the area of the exterior wall.
b. See Section 704.7 of International Building Code 2006 for unexposed surface temperature.
c. For occupancies in Group R-3, the maximum percentage of unprotected and protected exterior wall openings shall be 25 percent.
d. The area of openings in an open parking structure with a fire separation distance of greater than 10 feet shall not be limited.
e. For occupancies in Group H-2 or H-3, unprotected openings shall not be permitted for openings with a fire separation distance of 15 feet or less.
f. For requirements for fire walls for buildings with differing roof heights, see Section 705.6.1 of International Building Code 2006.
g. The area of unprotected and protected openings is not limited for occupancies in Group R-3, with a fire separation distance greater than 5 feet.
h. For special requirements for Group U occupancies, see Section 406.1.2 of International Building Code 2006.
i. Buildings who exterior bearing wall, exterior nonbearing wall and exterior structural frame are not required to be fire-resistance rated shall be permitted to have   unlimited un protected openings.
j. Includes accessory buildings to Group R-3.

!Codealert

A horizontal assembly is a fire-resistance-rated floor or roof assembly of materials designed to restrict the spread of fire in which continuity is maintained.

tected opening with a fire-protection rating of no less than 3/4 hour. The openings must also be vertically separated at least 3 feet (9.1 m) by spandrel girders or outside walls that have a 1-hour fire-resistance-rating. If not, you have to install flame barriers that extend at least 30 inches (76.2 cm) horizontally beyond the outside wall. Please note that whether you choose spandrel girders or flame barriers, a 1-hour fire-resistance rating is required. There are exceptions to this as follows:

- This section does not apply to buildings that are three stories or less in height.
- This section does not apply to buildings that are fully equipped with an automatic sprinkler system, but it must be throughout the whole building.
- This section does not apply to parking garages.

For buildings that are located on the same lot, opening protectives that have a fire-protection rating no less than 3/4 hour must be provided in every opening that is less than 15 feet (4.6 m) vertically above the roof of the adjoining building within 15 feet of the wall where the opening is located. But if the roof construction has a fire-resistance rating of no less than 1 hour with a minimum distance less than 10 feet (3.1 m), you can disregard the above. The 2006 International Building Code requires that all outside walls include parapets, which are walls placed at the edge of a roof. There are certain conditions in which a parapet does not need to be added to an outside wall:

- The wall is not required to be fire-resistance-rated because of fire-separation distance.
- The building's area is not more than 1,000 square feet (93 m^2) on any floor.
- Walls that end at roofs that are of less than 2-hour fire-resistance construction or if the roofs, including the decks and supports, are made entirely of noncombustible materials.
- One-hour fire-resistance walls that end at the underside of the roof, deck, or slab, if the following needs are met:
    - Roof and ceiling frames cannot be less than 1-hour fire-resistance-rated with a width of 4 feet (1,220 mm) for Groups R and U and 10 feet for all other occupancies.

- Roof and ceiling elements that are not parallel to the wall; the entire span of framing and supports cannot be less than 1-hour fire-resistance construction.
- Openings in the roof cannot be located within 5 (1,524 mm) feet of any 1-hour fire-resistance outside wall for Groups R and U and 10 feet for all other occupancies.
- The entire building must be provided with a roof covering that is no less than Class B.

• Groups R-2 and R-3 buildings that are covered with a Class C roof covering must have outside walls that end at the underside of the roof sheathing in Type III, IV, and V construction provided that the following conditions are met:

- The roof or deck is constructed of noncombustible materials that are approved or made of fire-retardant wood with a distance of 4 feet (1,220 mm) from the wall.
- The roof is protected with 0.625-inch (16 mm) Type X gypsum board directly beneath the roof or deck. The supports of the roof must have a minimum of 2-inch ledgers attached to the sides of the roof framing with a distance of 4 feet (1,220 mm) from the wall.
- The wall is allowed to have at least 25 percent of the outside walls containing unprotected openings.

PARAPETS

Now that you are aware of the conditions in which a parapet has to be used, you must also be aware of its construction requirements. Parapets must have the same fire-resistance rating as the other construction elements of an outside wall. They also must be noncombustible from 18 inches (457 mm) from the top downward, including counterflashing and coping materials. The code also requires that the parapet is not less than 30 inches (762 mm) above the point of the roof and wall intersection. And the parapet must extend to the same height of the roof within fire-separation distance, but in no case should the height be more than 30 inches (762 mm). As with any rule or code in the 2006 International Building Code, if you have any questions, please consult a building official.

WINDOWS

Any windows in outside walls that are required to have protected openings must comply with this chapter's provisions; other openings have to be protected with fire doors. Opening protectives are not required if the building is protected throughout with automatic sprinkler systems or an approved water curtains, and joints that are made in outside walls are allowed to have unprotected openings.

I have only begun to touch the surface of fire-resistance-rated construction. I know that you may be thinking that you already know this, but you must be prepared at all times for expected or even surprise visits from the building officer. Will you be prepared for this visit? Do your outside walls meet code? Have you installed automatic sprinkler systems throughout or where they are required? I am only trying to help you realize how seriously you must take this code. I want your building or structure to pass with flying colors, and it will if you follow the 2006 International Building Code.

FIRE WALLS

Now I am going to discuss the regulations for fire walls. Each portion of your building that is separated by one or more fire walls must be considered as a separate building. The extent and the location of the fire walls have to provide a complete separation. Wherever a fire wall separates occupancies, you must follow the strictest requirements. This includes any wall located on a lot line between buildings that are adjacent to each other or adapted for joint service,. Fire walls have the same conditions as the rest of the building in regard to duration. All fire walls, with the exception of Type V construction, have to be made of any approved noncombustible materials. I have included a fire-wall fire-resistance rating table (Table 7.2).

Fire walls must be continuous from outside to inside walls and must extend at least 18 inches (45.7 cm) beyond the outside surface walls. Exceptions to this rule are found below:

- Fire walls are permitted to end at the inside surface of combustible outside siding provided that the outside wall has a fire-resistance rating of at least 1 hour for a horizontal distance of at least 4 feet (1,220 mm) on both sides. Any openings in outside walls have to be protected by opening protectives that have a rating not less than 3/4 hour.

- Fire walls are permitted to end at the inside surface of noncombustible outside casing where the building on each side of the fire wall is protected by an automatic sprinkler system.

!Codealert

Mineral fiber is a type of insulation that is composed principally of fibers manufactured from rock, slag, or glass, with or without binders.

TABLE 7.2 Fire wall fire-resistance ratings.

GROUP	FIRE-RESISTANCE RATING (hours)
A, B, E, H-4, I, R-1, R-2, U	3[a]
F-1, H-3[b], H-5, M, S-1	3
H-1, H-2	4[b]
F-2, S-2, R-3, R-4	2

Exterior or outside walls at the point of intersection must have, on both sides, a 1-hour fire rating with a 3/4-hour opening protection. The rating must extend a minimum of 4 feet (1,220 mm) on each side starting from the intersection to the outside wall. If such walls form an angle equal to or greater than 180 degrees (3.14 rad), they do not need outside wall protection. Fire walls must extend to the outer edge of horizontal projections such as balconies, roof overhangs, canopies, and other similar structures that are within 4 feet (1,220 mm) of the fire wall. There are a few exceptions in regard to these projections:

• Horizontal projections without concealed spaces are exempt, provided that the outside wall behind and below the projection has no less than a 1-hour fire rating for a distance that is not less than the depth of the projection on either side of the fire wall. These openings have to have a fire protection rating of 3/4 hour, nothing less.

• Noncombustible horizontal projections with concealed spaces providing a minimum 1-hour-rated construction and extending through the concealed space are exempt. The wall is not required to extend under the projection and the openings within the outside walls must be protected by opening protectives that have a fire rating of 3/4 hour, nothing less.

• For combustible horizontal projections with concealed spaces, the fire wall only needs to extend through the concealed space to the outer edges of the projections. Openings within exterior walls must be protected by opening protectives not less than a 3/4 hour fire-protection rating.

!Codealert

Mineral wool is a synthetic vitreous fiber insulation made by melting predominantly igneous rock or furnace slag and other inorganic materials and then physically forming the melt into fibers.

Fire walls must have vertical continuity, meaning that they have to extend from the foundation to an ending point at least 30 inches (76.2 cm) above both adjacent roofs; however, 2-hour fire-rated walls are permitted to end at the underside of the roof, deck, or slab if the lower roof within 4 feet (1.2 m) of the wall has a 1-hour fire rating. Openings in the roof are not allowed to be located within 4 feet (1,1220 mm) of the fire wall, and each building must have no less than a Class B roof covering. Walls will be permitted to end at the underside of noncombustible roofs, decks, or slabs provided that both buildings have no less than a Class B roof covering. In buildings of Type III, IV, and V construction, walls are allowed to end at the underside of the roof or deck if there are no openings in the roof within 4 feet of the fire wall, the roof is covered with a minimum Class B covering, and the roof is either constructed of fire-retardant-treated wood with a distance of 4 feet (1,220 mm) on both sides of the wall or is protected with Type X gypsum that is 5/8 inch (15.9 mm) thick. If the building is located above a parking garage, its fire walls may be located above the garage from the horizontal separation between the garage and the building.

You must take note that when a fire wall serves as an outside wall for a building or for separated buildings that have different roof levels, the wall must end at a point no less than 30 inches (76.2 cm) above the lower roof level and the outside wall has to be at least 15 feet (4.6 m) above the lower roof. The wall must have a 1-hour fire rating on both sides of the openings protected by fire assemblies, which must be protected by no less than a 3/4 hour rating.

If the fire wall ends at the base of the roof, the lower roof within 10 feet (3 m) of the wall has to have no less than a 1-hour fire rating and the entire length and span of the supports for the roof assembly cannot have a rating of less than 1 hour. The lower roof cannot have openings that are located within 10 feet (3 m) of the fire wall.

Each opening through a fire wall must not be more than 120 square feet (11 m²), unless both buildings have an automatic sprinkler system throughout.

!Codealert

Windows in exterior walls required to have protected openings in accordance with other sections of the code or determined to be protected in accordance with Section 704.3 or 704.8 shall comply with Section 715.5. Other openings required to be protected with fire-door or shutter assemblies in accordance with other sections of this code or determined to be protected in accordance with Section 704.3 or 704.8 shall comply with Section 715.4.

The total width of openings at any floor level cannot be over 25 percent of the length of the wall, and openings are not allowed in party walls. Penetrations of fire walls and joints made in or between fire walls must comply with the provisions of this chapter, and, remember; ducts and air-transfer openings cannot penetrate fire walls unless they are not on a lot line and comply with this chapter. It is very important that you realize that fire walls are not to be confused with fire barriers, which I will talk about in the next section.

FIRE BARRIERS AND FIRE WINDOWS

Fire barriers must be made of materials permitted by the building type of construction, and fire-resistance-rated glazing, when tested in accordance with ASTM E 119, must also comply with this section. They have to be labeled or show some type of identification provided by an approved agency that shows the name of the manufacturer, the test standard, and the identifier that includes the fire-resistance rating. Exit enclosures, exit passageways, and horizontal exits must comply with Chapter 10 of the code book.

The fire barrier, horizontal assembly, or both separating a single occupancy into different areas must have a fire rating no less than indicated in Table 7.3.

When outside walls are part of a required fire-rated shaft or exit enclosure, they have to be in accordance with Chapter 10 provisions. Fire barriers must extend from the top of the floor/ceiling assembly to the bottom of the floor, roof slab, or deck above and securely attached.

Fire barriers must also be continuous through concealed spaces, such as the space above a suspended ceiling. All supporting construction must be protected to preserve the required fire rating of the fire-barrier supports. Hollow vertical spaces within a fire barrier must be fire-blocked at every floor level except that shaft enclosures are allowed to end at a top enclosure when in compliance with this section.

TABLE 7.3 Fire-resistance rating requirements for fire barrier assemblies between fire areas.

OCCUPANCY GROUP	FIRE-RESISTANCE RATING (hours)
H-1, H-2	4
F-1, H-3, S-1	3
A, B, E, F-2, H-4, H-5 I, M, R, S-2	2
U	1

!Codealert

The fire-resistance rating of the fire barrier separating atriums shall comply with Section 404.5.

All outside walls that are used as a part of a required fire-rated enclosure or separation have to comply with this section except for walls that are in accordance with Chapter 10. Openings in a fire barrier must be protected and are limited to a maximum aggregate width of 25 percent of the length of the wall, with the maximum area of any single opening not more than 156 square feet (15 m²). This does not apply to openings that have adjoining fire sprinklers throughout and whose assembly has been tested with ASTM E 119, governing fire doors serving as an exit.

Fire windows that are permitted in atrium separation walls are not limited to the maximum width of 25 percent. All penetrations of fire barriers must comply with this section. Penetrations into an exit enclosure or passageway will be allowed but only when permitted by Chapter 10. Joints, ducts, and air-transfer also openings comply with this section.

SHAFT ENCLOSURES

In this next section I will cover the code requirements for shaft enclosures. Shaft enclosures are required to protect openings and penetrations through floor/ceiling and roof/ceiling assemblies. Shaft enclosures must be constructed as fire barriers in accordance with this section, as horizontal assemblies, or both. Openings through a floor/ceiling assembly must be protected by a shaft enclosure. The following are exceptions regarding shaft enclosures required.

A shaft enclosure is not required for openings that are within (totally) individual residential dwellings and that connect four stories or less.

A shaft enclosure is not required in a building that is thoroughly equipped with an automatic sprinkler system for an escalator opening or stairway that is not a portion of a means of exit in compliance with Chapter 9, and if the area of the floor opening between stories does not exceed twice the horizontal projected area of the escalator or stairway and the opening is protected by a draft curtain and closely spaced sprinklers in accordance with NFPA 13. In groups other than Groups B and M, this exception is limited to openings that do not connect more than four stories.

Automatic shutters that protect a shaft opening must be noncombustible and have a fire rating of no less than 1 1/2 hours. These shutters must be designed so that they close and shut off immediately upon the detection of smoke. Automatic shutters cannot operate at a speed of less than 30 feet per minute (152.4 mm/s), and they need to have a sensitive leading edge to stop movement when they come into contact with any obstacle and proceed when the obstacle is removed from their path.

In groups other than H occupancies, a shaft enclosure is not required for floor openings that comply with atrium requirements in Chapter 4. In groups other than I-2 and I-3, a shaft enclosure is not required for a floor opening or an air-transfer opening that complies with the following:

• Does not connect more than two stories

• Is not part of the required means of exit, except for Chapter 10 provisions

• Is not concealed within the building construction

• Is not open to a corridor in Group I and R occupancies or on floors that do not have sprinklers

• Is separated from floor openings and air-transfer openings serving other floors by construction conforming to code

A shaft enclosure is not required for the following:

• Automobile ramps in open and closed parking garages constructed in accordance with Chapter 4

• Floor openings between a mezzanine and the floor below

• Joints protected by a fire-resistant joint system

• Floor openings created by unenclosed stairs or ramps

• Floor openings protected by floor fire doors

• Penetrations by a pipe, tube, conduit, wire, cable, duct, or vent that is protected under the requirements of any other sections of this code

Shaft enclosures must have a fire-resistance rating of not less than 2 hours in connections of four stories or more but no less than 1 hour for less than four stories. The number of stories can include basements but not mezzanines. They must

!Codealert
The fire barrier separating incidental-use areas shall have a fire-resistance rating of not less than that indicated in Table 508.2 of the code.

!Codealert

Fire barriers separating control areas shall have a fire-resistance rating of not less than that required in Section 414.2.3.

also have a fire-resistance rating no less than that of the floor assembly penetrated but not higher than 2 hours.

All shaft enclosures must be made as fire barriers, horizontal assemblies, or both. If your outside walls are part of the required shaft enclosure, they must comply with the section regarding outside walls and the fire-rating enclosure requirements do not apply except for walls that are required to be protected in accordance with Chapter 10 provisions for exterior exit ramps and stairways.

Openings in a shaft enclosure must be tested in accordance with this chapter as required for fire barriers; all doors must be self- or automatic-closing. Any opening other than those that are necessary for the purpose of the shaft cannot be used in shaft enclosures. This applies to penetrations as well; any penetration of a shaft enclosure by ducts and air-transfer openings must comply with this chapter.

Some shafts do not go all the way to the bottom of the building or structure. When this happens shafts must be enclosed at the lowest level with construction of the same fire rating as the lowest floor level where the shaft passes through. They also must end in a room that has uses that are related to the purpose of the shaft. This room must be separated from the rest of the building by a fire barrier that is rated and includes opening protectives that are at least the same as the protection required for the shaft enclosure.

Shafts that do not extend to the bottom of the building or structure must be protected by approved fire dampers as well, unless there are no openings in the shaft except for the bottom. If this is the case, the bottom of the shaft must be closed off around the penetrating objects to prevent drafts.

Shaft enclosures that contain a refuse or laundry chute cannot be used for any other purpose. The fire-resistance rating and the protection at the bottom of the shaft are not required if there are no combustibles in the shaft and no openings or other penetrations through the shaft enclosure to the inside of the building. A shaft enclosure that does not extend to the bottom of the roof, deck, or slab of the building must be enclosed at the top with construction of the same fire rating as the top floor that is penetrated by the shaft and not any less.

All refuse and laundry chutes, access and end rooms, and incinerator rooms must meet certain requirements. This list of requirements does not apply to chutes that are used and contained within a single dwelling:

• Openings into chutes cannot be located in corridors or hallways.

• Doors must be self- or automatic-closing when the automatic smoke detector goes off; this does not include heat-activated closing devices.

• All shaft enclosures that contain a laundry chute must be constructed of materials allowed for the type of building construction.

• Access openings must be located in rooms that are enclosed by a fire barrier and that have a fire-resistance rating of no less than 1 hour. These openings must be protected by protectives that have a fire rating of no less than 3/4 hour. All doors must be self- or automatic-closing when smoke is detected.

• Laundry chutes must end in an enclosed room that is separated from the rest of the building by a fire barrier that has a fire rating of no less than 1 hour. Openings must have a fire rating of no less than 3/4 hour, and doors must be automatic- or self-closing.

• Refuse chutes cannot terminate in an incinerator room.

• Laundry rooms that do not have chutes are not required to comply with the above.

OTHER TYPES OF SHAFTS

Chutes are not the only shafts in a building. Elevators, dumbwaiters, and other hoistways are other means in which a shaft can be used. An enclosed lobby must be provided at each floor where an elevator-shaft enclosure connects more than three stories. The lobby is used to separate the elevator-shaft-enclosure doors from each floor by fire partitions equal to the fire rating of the hall and the required

!Codealert

Where exterior walls serve as a part of a required fire-resistance-rated enclosure or separation, such walls shall comply with the requirements of Section 704 for exterior walls, and the fire-resistance enclosure or separation requirements shall not apply.

opening protection. Elevator lobbies must have at least one means of exit that complies with Chapter 10. Enclosed elevator lobbies are not required at the street level if the entire level is equipped with an automatic sprinkler system. All additional doors have to be tested in accordance with UL 1784 without an artificial bottom seal.

In groups other than Group I-3 and buildings that have occupied floors that are more than 74 feet (23 m) above the lowest level of fire-department access, enclosed elevator lobbies are not required. Smoke partitions can be permitted instead of fire partitions to separate the lobby at each floor as long as there is an automatic sprinkler system installed.

If the elevator hoistway is pressurized, elevator lobbies are not required. When elevator-hoistway pressurization is provided instead of enclosed lobbies, the pressurization system must comply with the provisions of this section. Elevator hoistways must maintains a minimum positive pressure of 0.04 inches of water column and maximum positive pressure of 0.06 inches of water column with respect to adjacent occupied space on all floors. The pressure has to be measured at the midpoint of each hoistway door with all the hoistway doors on the ground level closed. Air intake must be taken from an outside, uncontaminated source with a distance of 20 feet (6.1 m) from any air exhaust. Any duct that is part of the pressurization system must be protected with the same fire-resistance rating as is required for the elevator-shaft enclosure.

Any fan that is provided for the pressurization system is required to comply with the following section. A fan system that is located within the building must be protected with the same fire-resistance rating that is required for the elevator-shaft enclosure. The fan system must also be equipped with a smoke detector that will automatically shut down the system when smoke is detected.

Each elevator is required to have a separate fan system. A fan system can either be adjustable with a capacity of at least 1,000 cfm (.4719 m3/S) per door or specified by a registered design professional to meet the requirements of a pressurization system.

All pressurized systems need to have self-activating fire-alarm and smoke-detecting systems.

!Codealert
Penetrations in a fire barrier by ducts and air-transfer openings shall comply with Section 716.

!Codealert

Identification for fire-protection-rated glazing has current updates, which can be found in Section 715.6.3.1 of the code.

ings other than Type IIB, IIIB, or VB construction. Any smoke barrier with penetrations, joints, ducts, and air-transfer openings must comply with the provisions of this chapter.

Smoke partitions are not required to have a fire-resistance rating unless required to do so elsewhere in the code. Smoke partitions extend from the top of the foundation or floor to the bottom of the roof to limit the transfer of smoke. All windows and doors, but not including louvers, must be tested in accordance with UL 1784 before they are considered to be resistant to the free passage of smoke. Any free spaces around penetrating items and in joints must be filled with an approved material to limit the free passage of smoke; this is also true for ducts and air-transfer openings, unless installation interferes with the operation of a required smoke-control system.

HORIZONTAL ASSEMBLIES

A horizontal assembly refers to roofs and floors. Floor and roof assemblies that are required to have a fire-resistance rating must comply with this section. All floor and roof assemblies must be made of materials allowed by the type of building construction, and the fire-resistance rating cannot be less than that required by the building itself. But what if you have floor assemblies that separate mixed occupancies? In such cases the assembly must have a fire-resistance rating no less than that required by Chapter 5, based on the occupancies that are being separated.

If a floor assembly separates a single occupancy into different fire areas, the assembly must have a fire-resistance rating of no less than that discussed earlier in this chapter. All floor assemblies that separate living spaces in the same building or sleeping units in occupancies in Groups R-1 hotel occupancies, R-2, and I-1 must be a minimum of 1-hour fire-resistance-rated construction. The only exception to this are dwelling and sleeping units that are used as separation in buildings of Type IIB, IIIB, and VB construction, which must have a fire-resistance rating of no less than 1/2 hour in buildings that are equipped throughout with an automatic sprinkler system in accordance with Chapter 9.

Ceiling Panels

Where the weight of lay-in ceiling panels that are used as part of a fire-resistance-rated floor/ceiling or roof/ceiling assembly is not strong enough to resist an upward force of 1 pound/foot^2 (48 Pa), you must install wire or another approved device to prevent vertical bowing under such upward force. Access doors are permitted in ceilings of fire-resistance-rated assemblies provided that the doors are tested in accordance with ASTM E 119 as horizontal assemblies and labeled as such. In 1-hour-fire-resistance-rated floor and roof construction neither the ceiling nor the floor covering is required to be installed where unusable crawl or attic space occurs. All assemblies must be continuous without openings, penetrations, or joints except where allowed elsewhere in this chapter and in Chapter 10.

Skylights

Skylights and other penetrations through a fire-resistance-rated roof deck or slab are permitted to be unprotected, provided that their structural reliability is continuously maintained. Take note that unprotected skylights are not permitted under this section. Penetrations, joints, ducts, and air-transfer openings must comply with previous sections of this chapter. Floor fire-door assemblies that are used to protect openings in fire-resistance-rated floors must be tested in accordance with NFPA 288, and must have a fire-resistance rating no less than the assembly being penetrated. All floor fire doors must be labeled by an approved agency, be permanently attached, and include the name of the manufacturer, the test standard, and the fire-resistance rating.

PENETRATIONS

This section concerns materials and methods of construction used to protect through penetrations, which are openings that pass through entire assemblies and casings of horizontal assemblies and fire-resistance-rated wall assemblies. Where

---

!Codealert

When dealing with the fire-resistance rating of structural members, the king studs and boundary elements that are integral elements in load-bearing walls of light-framed construction shall be permitted to have required fire-resistance ratings provided by the membrane protection provided for the load-bearing wall.

!Codealert

Alternative methods for determining fire-protection ratings are new and can be found in Section 715.3 of the code.

sleeves are used, they must be securely fastened to the assembly that is penetrated. The space between the item contained in the sleeve and the sleeve itself and any space between the sleeve and the assembly penetrated must be protected. Insulation and coverings on or in the penetrating item must not penetrate the assembly unless the specific material used has been tested as part of the assembly.

Penetration into or through fire walls, fire-barrier walls, smoke-barrier walls and fire partitions must comply with the following:

• In concrete or masonry walls where the penetrating item has a maximum 6-inch (15.2 cm) nominal diameter and the area of the opening through the wall is not more than 144 square inches (0.0929 m^2), concrete, grout, or mortar is allowed if it is installed at the full thickness required to maintain the fire-resistance rating or if the material used to fill the annular space will prevent the passage of flame and hot gases that can ignite cotton waste when put up against ASTM E 119 time-temperature fire conditions under a minimum positive pressure difference of 0.01 inch (2.49 Pa) of water at the location of the penetration for the time period that is equal to the fire-resistance rating of the construction penetrated.

• All penetrations must be installed as tested in an approved fire-resistance-rated assembly.

• Penetrations must be protected by an approved firestop system installed and tested in accordance with ASTM E 814 or UL 1479, with a minimum positive pressure differential of 0.01 inch (2.49 Pa) of water, and must have an F rating of no less that of the fire-resistance rating.

• Where walls or partitions are required to have a fire-resistance rating, recessed fixtures cannot be installed in a way that will reduce the required fire resistance, except for the following: membrane penetrations of maximum 2-hour fire-resistance-rated walls and partitions by steel electrical boxes that are not more than 16 square inches (0.0103 m^2) in area are allowed but only if the total area of the openings through the membrane is not more than 100 square inches (0.0645 m^2) in any 100 square feet of wall area. The annular space between the wall membranes and the box cannot be more than 1/8 inch (0.31 cm).

Steel electrical boxes on opposite sides of walls or partitions must be separated by one of the following:

- By a horizontal distance of no less than 24 inches (61 cm)

- By a horizontal distance of not less than the depth of a hole in the wall filled with cellulose loose-fill, rock wool, or slag mineral wool insulation

- By solid fire blocking

- By listed putty pads

- By any other listed materials and methods

Membrane penetrations by listed electrical boxes of any material, and the boxes have been tested for use in fire-resistance-rated assemblies must be installed in accordance with the instructions that are included in the listing. The space between the wall and the box cannot be more than 1/8 inch (3.1 mm) unless listed otherwise. Boxes on opposite sides of a wall or partition must be separated according to the requirements of the list above, and the annular space created by the penetration of a fire sprinkler does not require recessed fixtures, but only if it is covered by a metal escutcheon plate. Dissimilar materials such as noncombustible penetrating items must not connect to combustible items beyond the point of fire stopping unless it can be demonstrated that the fire-resistance integrity of the wall is maintained.

FIRE-RESISTANT JOINT SYSTEMS

Joints that are installed in or between fire-resistance-rated walls, floor or floor/ceiling assemblies, and roofs or roof/ceiling assemblies must be protected by an approved fire-resistant joint system designed to resist the passage of fire for a time period that is not less than the required rating for the other elements. The void or space created at the intersection of a floor/ceiling assembly and an outside curtain-wall assembly must be protected in accordance with this section. Fire-resistant joint systems are not required for joints in the following locations:

- Floors within a single dwelling unit
- Floors where the joint is protected by a shaft enclosure

!Definitionalert

Smoke Damper: A listed device installed in ducts and air-transfer openings designed to resist the passage of smoke. The device is installed to operate automatically, controlled by a smoke-detection system, and, where required, capable of being positioned from a fire command center.

!Codealert

See Section 715.4.3.2 of the code for updates on glazing in door assemblies.

• Floors within atriums where the space adjacent to the atrium is included in the volume of the atrium for smoke-control purposes

• Floors within malls

• Floors within open parking structures

• Mezzanine floors

• Walls that are allowed to have unprotected openings

• Roofs where openings are permitted

• Control joints that are not more than a maximum width of 0.625 inch (1.59 cm) and tested in accordance with ASTM E 119

Fire-resistant joint systems must be securely installed in or on the joint for its entire length so that it will not dislodge, loosen, or otherwise impair its ability to accommodate expected building movements and so that it will resist the passage of fire and hot gases. Fire-resistant joint systems must be tested in accordance with the requirements of either ASTM E 1966 or UL 2079. Nonsymmetrical wall joint systems must be tested with both faces exposed to the furnace, and the assigned fire-resistance rating must be the shortest time from the two tests.

When evidence is furnished to show that a wall was tested with the least fire-resistant side exposed to the furnace and the building official agrees, the wall does not need to be subjected to tests from the opposite side. For outside walls with a horizontal fire-separation distance more than 5 feet (1.5 m), the joint system must be tested for inside fire exposure only.

Where fire resistance-rated floor or floor/ceiling assemblies are required, empty spaces created at the intersection of the outside curtain-wall assemblies and such floor assemblies must be sealed with an approved material or system to prevent fire from spreading inside. These materials or systems must be securely installed and capable of preventing the passage of flame and hot gases that are hot enough to ignite cotton waste when subjected either to ASTM E 119 time-temperature fire conditions under a minimum positive pressure differential of 0.01 inch (0.254 mm) of water column (2.5 Pa) or tested in accordance with ASTM E 2307 for the time period at least equal to the fire-resistance rating of the floor assembly.

Height and fire-resistance requirements for curtain-wall spandrels must comply with the section of this chapter that does not require a fire-resistance-rated spandrel wall. Fire-resistant joint systems in smoke barriers must be tested in accordance with the requirements of UL 2079 for air leakage. The air-leakage rate of the joint must not be more than 5 cfm per lineal foot (0.00775 m3/slm) of joint at 0.30 inch (7.47 Pa) of water for both the immediate surrounding-area temperature and elevated temperatures.

STRUCTURAL MEMBERS

The fire-resistance rating of structural members and assemblies must comply with the requirements for the type of construction and cannot be less than the rating required for the fire-resistance-rated assemblies supported except for fire barriers, partitions, and smoke barriers as described earlier in this chapter. Protection of columns, girders, trusses, beams, lintels, or other structural pieces that are required to have a fire-resistance rating must also comply with this section. Any that are required to have a fire-resistance rating and that support more than two floors, one floor and a roof, a load-bearing wall, or a non-load-bearing wall more than two stories high must be individually protected on all sides for the full length with materials with the required fire-resistance rating.

Other structural pieces required to have a fire-resistant membrane must have individual encasement by a covering or ceiling protection as specified earlier or by a combination of both. King studs and boundary elements that are integral parts of load-bearing walls in light-framed construction must have fire-resistance ratings equivalent to the covered protection provided for the load-bearing wall. Where columns need a fire-resistance rating, the entire column, including its connections to beams or girders, must be protected.

Where a column extends through a ceiling, the fire resistance of the column must be continuous from the top of the foundation or floor/ceiling assembly below through the ceiling space to the top of the column. The required thickness and

!**Code**alert

Smoke dampers are not required at penetrations of shafts where ducts are used as part of an approved mechanical smoke-control system designed in accordance with Section 909 and where the smoke damper will interfere with the operation of the smoke-control system.

!Definitionalert

Fire Partition: A vertical assembly of materials designed to restrict the spread of fire and in which openings are protected.

construction of fire-resistance-rated assemblies enclosing trusses must be based on the results of full-scale tests or combinations of tests on truss components or on approved calculations based on such tests that satisfactorily demonstrate that the assembly has the required fire resistance. The edges of lugs, brackets, rivets, and bolt heads that are attached to structural members are allowed to exceed 1 inch (2.5 cm) of the surface of the fire protection.

The thickness of protection for concrete or masonry reinforcement must be measured to the outside of the reinforcement except that stirrups and spiral reinforcement ties are allowed to project no more than 0.5 inch (1.3 cm) into the protection. Any and all pipes, wires, conduits, ducts, or other service facilities must not be embedded in the required fire-protective covering of a structural member that is required to be individually encased.

Anywhere the fire-protective covering of a structural member is subject to impact damage from moving vehicles, the handling of merchandise, or other activity, the fire protective covering must be protected by corner guards or by a substantial jacket of metal or other noncombustible material to a height that will provide full protection but no less than 5 feet (1.5 m) from the finished floor.

Exterior Structural Members

Exterior structural members that are located within the outside walls or on the outside of a building or structure must be provided with the highest fire-resistance rating as determined by Table 6.1 for the type of building element based on the type

!Codealert

Section 716.6.2 of the code deals with membrane penetrations and contains numerous changes in the code language.

of construction of the building. For outside bearing walls see Table 6.2 for calculations based on the fire-separation distance. These tables can be found in Chapter 6 of this book.

Fire protection is not required at the bottom flange of lintels, shelf angles, and plates spanning no more than 6 feet (1.8 m), whether part of the structural frame or not, and is also not required from the bottom flange or any part of the structural frame regardless of span.

Fire-resistance ratings for the isolation system must meet the fire-resistance rating required for columns, walls, or other structural elements in which the isolation system is installed in accordance with Table 6.1. Isolation systems that are required to have a fire-resistance rating must be protected with approved materials or construction assemblies designed to provide the same degree of fire resistance as the structural element in which it is installed when tested in accordance with ASTM E 119. This isolation protection system must be capable of preventing the transfer of heat to the isolation unit in such a manner that the required gravity load-carrying capacity of the isolator unit will not be damaged after exposure to the standard time-temperature-curve fire test in ASTM E 119 for no less time than required for the fire-resistance rating of the structural element of installation. This isolation-system protection applied to isolator units must be correctly designed and securely installed so that it does not dislodge, loosen, become damaged, or harm its ability to accommodate the seismic movements for which it is designed.

OPENING PROTECTIVES

Any opening protectives that are required by other sections of this code must comply with the provisions of this section. However, labeled fire-resistance-rated glazing tested as part of a wall assembly in accordance with ASTM E 119 does not need to comply with this section. There are alternative methods for determining fire-protection ratings, but they must be based on the fire exposure and acceptance criteria specified in NFPA 252 or NFPA 257. The required fire resistance of an opening protective can be recognized by any of these methods:

!Definitionalert

Ceiling Radiation Damper: A listed device installed in a ceiling membrane of a fire-resistance-rated floor/ceiling or roof/ceiling assembly to limit automatically the radiative heat transfer through an air inlet/outlet opening.

TABLE 7.5 Fire window assembly fire protection ratings.

TYPE OF ASSEMBLY	REQUIRED ASSEMBLY RATING (hours)	MINIMUM FIRE WINDOW ASSEMBLY RATING (hours)
Interior walls:		
Fire walls	All	NP[a]
Fire barriers	>1	NP[a]
Smoke barriers	1	$3/4$
and fire partitions	1	$3/4$
Exterior walls	>1	$1\,1/2$
	1	$3/4$
Party wall	All	NP

NP = Not Permitted.

a. Not permitted except as specified in Section 715/2 of *International Building Code 2006.*

side, and they have to be identified by a special mark or letter that has to be at least 6 inches (15.2 cm) high. If your fire shutters are the rolling type, they have to be automatic-closing and approved.

DUCTS AND AIR TRANSFER OPENINGS

This section will explain the protection requirements of protection for duct penetrations and air-transfer openings in assemblies. When you or a contractor installs fire, smoke, combination fire/smoke, and ceiling radiation dampers that are within air-distribution and smoke-control systems, they must be installed in accordance with the provisions of this section of the code. The manufacturer's installation instructions and the dampers' listing must always be followed. If the installation of a fire damper should interfere with the use of a smoke-control system, you must use an alternative protection, but remember that the alternate has to be approved for use.

All fire dampers used for hazardous exhaust duct systems must comply with the International Mechanical Code. Assemblies with less than a 3-hour fire-resistance-rating must have a minimum of a 1 1/2-hour damper rating. And assemblies that have 3-hour or higher ratings must have a minimum of a 3-hour damper rating. The fire-damper activating device must meet one of the following requirements:

• The operating system must be approximately 50 degrees F (10 degrees C) above the normal temperature within the duct system but no less than 160 degrees F (71 degrees C).

> # !Definitionalert
>
> **Ceramic Fiber Blanket:** A mineral-wool insulation material made of aluminum-silica fibers and weighing 4 to 10 pounds per cubic foot (pcf) (64 to 160 kg/m^3).

- The operating temperature must be no more than 286 degrees F (141 degrees C) within a smoke-control system that complies with Chapter 9.
- If a combination fire/smoke damper is located in a smoke-control system complying with Chapter 9, the operating temperature rating must be approximately 50 degrees F (10 degrees C) above the minimum smoke-control system activation temperature up to maximum temperature of 350 degrees F (177 degrees C). The temperature must not be higher than the UL 555S degradation-test temperature rating for a combination fire/smoke damper.

Smoke Dampers

Smoke-damper leakage ratings cannot be less than Class II, and elevated temperature ratings cannot be less than 250 degrees F (121 degrees C). Smoke dampers must close when a smoke detector(s) is activated. If a damper is installed within a duct, a smoke detector must also be installed within 5 feet (1.5 m) of the damper with no air outlets or inlets in between. If a damper is installed above a smoke-barrier door in a smoke barrier, a spot-type detector listed for releasing service must be installed on either side of the smoke-barrier-door opening. If a damper is installed within an unducted opening in a wall, a spot-type detector listed for releasing service must be installed within 5 feet (1.5 m) horizontally of the damper.

If a damper is installed in a corridor wall or ceiling, the damper has to be controlled by a smoke detector system. If a total-coverage smoke-detector system is provided within areas using a heating, ventilation, and air-conditioning (HVAC) system, dampers must be controlled by the smoke-detection system. Fire and smoke dampers must be provided with an approved means of access large enough to permit inspections and maintenance of the damper and its parts. The access cannot affect the reliability of fire-resistance-rated assemblies. Access points must have a label that reads: Fire/Smoke Damper, Smoke Damper, or Fire Damper. The letters in the label cannot be any less than 0.5 inch (1.27 cm) high.

The access doors in ducts have to fit tight and be suitable for the required duct construction. Fire, smoke, combination fire/smoke, and ceiling radiation dampers must be provided for fire walls in accordance with this chapter, for fire barriers in accordance with Chapter 10, and in shaft enclosures except where steel exhaust subducts are extended at least 22 inches (5.6 cm) vertically in exhaust shafts, provided that there is a continuous airflow upward to the outside or where penetrations are tested in accordance with ASTM E 119.

Fire dampers are also not required at penetrations of shafts where ducts are used as part of an approved smoke-control system designed and installed in accordance with Chapter 9, where the fire damper will interfere with the operation of the smoke-control system, or where the penetrations are in parking garages or exhaust or supply shafts that are separated from other building shafts by no less than a 2-hour fire-resistance-rated construction.

Groups B and R occupancies that are thoroughly equipped with an automatic sprinkler system in accordance with Chapter 9 are not required to have smoke dampers or shafts if kitchen, clothes-dryer, and bathroom exhaust openings are installed with steel exhaust subducts that have a wall at least 0.019 inches (0.48 mm) thick and extend at least 22 inches (5.6 cm) vertically. An exhaust fan that is powered continuously must be installed at the upper end of the shaft in accordance with Chapter 9 to ensure that a continuous airflow is maintained.

Smoke dampers are not required at penetrations of exhaust or supply shafts in parking garages that are separated from other building shafts and where ducts are used as part of an approved mechanical smoke-control system. This brings us to the end of smoke dampers and to the beginning of fire partitions.

Fire Dampers

Ducts and air-transfer openings that penetrate fire partitions must be protected with listed fire dampers. Occupancies other than Group H do not require a fire partition if the following conditions are met:

- Buildings with tenant separations or corridor walls that are equipped throughout with an automatic sprinkler system and ducts protected as a through penetration
- Tenant partitions in covered mall buildings where the walls are not required by provisions elsewhere in the code to extend to the bottom of the floor or roof deck above
- The duct system is constructed of approved materials in accordance with the International Mechanical Code, and the duct that penetrates the wall meets the following requirements:
    - The duct cannot be more than 100 square inches (0.06 m^2).
    - The duct must be constructed of steel that is a minimum of 0.0217 inch (0.55 mm) thick.
    - The duct cannot have openings that share the corridor with adjacent spaces or rooms.
    - The duct cannot terminate as a wall register in the fire-resistance-rated wall.

- A minimum 12-inch-long (30.5 cm) by 0.060-inch-thick (1.52 mm) steel sleeve has to be centered in each duct opening.
- The sleeve must be secured to both sides of the wall and all four sides of the sleeve with steel retaining angles that are a minimum of 1.5 inches by 1.5 inches by 0.060-inch (38 mm by 38 mm by 1.52 mm).
- The retaining angles must be secured to the sleeve and the wall with No. 10 (M5) screws.
- The annular space between the steel sleeve and the wall opening must be filled with mineral wool batting on all sides.

Smoke Dampers

A listed smoke damper is designed to resist smoke passing through areas of buildings and must be provided at each point where a duct or air-transfer opening penetrates a corridor. However, please note that smoke dampers are not required when the building has an approved smoke-control system in place throughout or in corridor penetrations if the duct is constructed of steel that is no less than 0.019 inches (0.48 mm) thick and there are no openings in use for corridors.

You must also have a smoke damper in place at each duct or air-transfer opening penetrating a smoke barrier unless the openings in the duct are part of a single smoke compartment that is constructed of steel. Penetrations by ducts and air-transfer openings of a floor, floor/ceiling assembly, or ceiling covering of a roof/ceiling assembly must be protected by a shaft enclosure that complies with this chapter of the code. In occupancies other than Groups I-2 and I-3, a duct that is constructed of approved materials in accordance with the International Mechanical Code and that penetrates a fire-resistance-rated floor/ceiling that connects no more than two stories is permitted without shaft enclosure protection, provided that that a listed fire damper is installed at the floor line or the duct is protected.

A duct is permitted to penetrate three floors or less without a fire damper at each floor, provided it meets all the following requirements:

• The duct must be contained and located within the cavity of a wall and must be constructed of steel that is no less than 0.019 inches (0.48 mm) (26 gauge) in thickness.

• The duct must open into only one dwelling or sleeping unit, and the duct system must be continuous from the unit to the outside of the building.

• The duct cannot be more than 4 inches (10.2 cm) in nominal diameter, and the total area of the ducts cannot be more than 100 square inches (0.065 m^2) in any 100 square feet (9.3 m^2) of floor area.

• The annular space around the duct is protected with materials that prevent the passage of flame and hot gases sufficient to ignite cotton waste when exposed to ASTM E 119 time-temperature conditions under a minimum positive pressure difference of 0.01 inch (2.49 Pa) of water at the location of the penetration for a

flective-foil insulation that are installed behind and in contact with the unexposed surface of the ceiling, wall, or floor finish. However, exposed insulation that is installed on attic floors must have a critical radiant flux of no less than 0.12 watt per square centimeter when tested in accordance with ASTM E 970.

Another type of insulation is loose-fill. This type of insulation cannot be heaped in the ASTM E 84 apparatus without a screen or support unless it complies with the flame-spread and smoke-developed limits. If the loose-fill insulation that you are using is cellulose, it does not have to comply with the frame-spread index that is required of CAN/ULC S102.2. But you need to be aware that loose-fill insulation does have to comply with CPSC 16 CFR, Part 1209, and CPSC 16, Part 1404, when applicable and must be labeled as such. Combustible roof insulation that does not comply with this section can be used in any type of construction, but it needs to be covered with an approved roof covering.

PRESCRIPTIVE FIRE RESISTANCE

This section contains prescriptive details of fire-resistance-rated building elements, which include the thickness of protective coverings, masonry protection, and plaster application.

CALCULATED FIRE RESISTANCE

This section contains procedures or uses for fire-resistant materials that have been established by calculations. I have included some definition alerts for additional information.

Table 7.6 contains the minimum equivalent thicknesses of cast-in-place or pre-cast walls for fire-resistance ratings of 1 to 4 hours.

For hollow-core and precast-concrete wall panels in which the cores are of constant cross section throughout the length, you can calculate the equivalent thickness by dividing the net cross-sectional area (which is found by subtracting the gross cross panel from the area of the cores) of the panel by the equired width. Where all the core spaces of hollow-core wall panels are filled with loose-fill material, such as expanded shale, clay, slag, vermiculite, or perlite, the fire-resistance rating of the wall is the same as that of a solid wall of the same concrete type and the same overall thickness.

The thickness of the panels with tapered cross sections must be determined at a distance $2t$ (t represents the minimum thickness) or 6 inches (1.5 cm), whichever is less, from the point of minimum thickness. The equivalent thickness of panels with ribbed or undulating surfaces will be determined by one of the following expressions:

TABLE 7.6 Minimum equivalent thickness of cast-in-place or precast concrete walls, load-bearing or nonload-bearing.

CONCRETE TYPE	MINIMUM SLAB THICKNESS (inches) FOR FIRE-RESISTANCE RATING OF				
	1-hour	1 ½-hour	2-hour	3-hour	4-hour
Siliceous	3.5	4.3	5.0	6.2	7.0
Carbonate	3.2	4.0	4.6	5.7	6.6
Sand-Lightweight	2.7	3.3	3.8	4.6	5.4
Lightweight	2.5	3.1	3.6	4.4	5.1

For SI: 1 inch = 25.4 mm

For $s \geq 4t$, the thickness to be used must be t.

For $s \leq 2t$, the thickness to be used must be t.

For $4t > s > 2t$, the thickness to be used must be

$$t + \left(\frac{4t}{s} - 1 \right) \left(t_e - t \right)$$
(Equation 7.1)

where

s = Spacing of ribs or undulations

t = Minimum thickness

t_e = Equivalent thickness of the panel calculated as the cross-sectional area divided by the width, in which the maximum thickness used in the calculation cannot exceed $2t$

For walls that consist of two widths of different types of concrete, the fire-resistance ratings can be determined from Figure 7.2 and Equation 7.2.

$$R = (R_1^{0.59} + R_2^{0.59} + \ldots + R_n^{0.59})^{1.7}$$
(Equation 7.2)

where

R = Fire endurance of the assembly in minutes

R_1, R_2, and R_n = Fire endurances of the individual widths in minutes. Values of $R_n^{0.59}$ for use in Equation 7.1 are given in Table 7.7. Calculated fire-resistance ratings are shown in Table 7.7.

The fire-resistance ratings of precast-concrete wall panels that have a layer of foam-plastic insulation squashed between two widths of concrete are determined by the use of Equation 7.2. Foam-plastic insulation that has a total thickness of less than 1 inch (25 mm) may be ignored. The R_n value for thickness of foam-plastic insulation of 1 inch (25 mm) or greater, for use in the calculation, is 5 minutes; therefore R_n 0.59 = 2.5.

When you have joints that are between precast-concrete wall panels that are not insulated as required by this section, they will be considered as openings in walls. Un-insulated joints must be included in determining the percentage of openings that are allowed by Table 7.1.

I've discussed concrete wall panels and foam insulation and have provided several tables and equations for further information. Don't be fooled—there will be more. But first I want to go over nonexposed sides that gypsum wallboard or

TABLE 7.7 Fire-resistance ratings based on $R^{0.59}$.

OCCUPANCY GROUP	FIRE-RESISTANCE RATING (hours)
H-1, H-2	4
F-1, H-3, S-1	3
A, B, E, F-2, H-4, H-5 I, M, R, S-2	2
U	1

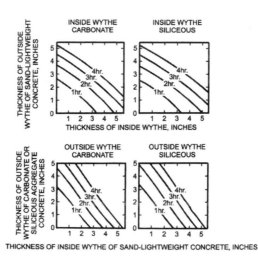

FIGURE 7.2 Fire-resistance ratings of two-wythe concrete walls.

plaster is applied to. When the finish of wallboard or plaster is applied to the side of the wall that is not exposed to the total fire-resistance rating, it has to be determined by the thickness of the finish. To determine the thickness, you must multiply the actual thickness by the factor that applies, based on the type of aggregate in the concrete. I have included another table that you can use as a guide in determining the thickness.

If you are applying gypsum wallboard or plaster to the fire-exposed side of the wall, the contribution of the finish to the total fire-resistance rating must be determined as well. You may be wondering about the walls that have no finish on one side or the other or that have different types of thicknesses or finishes on each side. If this is truly the case, the calculation procedures as discussed above must be done twice, assuming either side of the wall to be the fire-exposed side. The fire-resistance rating of the wall cannot be more than the lower of the two values.

If you have an outside wall with more than 5 feet (1.5 m) of horizontal separation, the fire must be assumed to occur on the inside only. When finishes that are applied to one or both sides of a concrete wall contribute to the fire-resistance rating, the concrete alone cannot provide less than half the total required fire-resistance rating. Also, please be aware that the input to the fire resistance of the finish on the non-fire-exposed side of the load-bearing wall cannot be more than half the input of the concrete alone.

Concrete floors and/or roof slabs also require a minimum thickness to comply with the code. Depending on the type of concrete that is used, the fire-resistance rating is from 1 to 4 hours. For example, if you were to use a lightweight sand concrete, the rating for 1 hour is 2.7; if you were to use a siliceous type of concrete, the rating for 3 hours is 6.2. Be sure to check with the 2006 International Building Code and your local building official if you are unsure of any of the fire-resistance ratings.

TABLE 7.8 Multiplying factor for finishes on nonfire-exposed side of wall.

TYPE OF FINISH APPLIED TO MASONRY WALL	TYPE OF AGGREGATE USED IN CONCRETE OR CONCRETE MASONRY			
	Concrete: siliceous or carbonate Masonry: siliceous or calcareous gravel	Concrete: sand lightweight concrete Masonry: limestone, cinders or unexpanded slag	Concrete: lightweight concrete Masonry: expanded shale, clay or slate	Concrete: pumice, or expanded slag
Portland cement-sand plaster	1.00	0.75[a]	0.75[a]	0.50[a]
Gypsum-sand plaster or gypsum wallboard	1.25	1.00	1.00	1.00
Gypsum-vermiculite or perlite plaster	1.75	1.50	1.50	1.25

For SI: 1 inch = 25.4 mm.
a. For portland cement-sand plaster 5/8 inch or less in thickness and applied directly to the masonry on the nonfire-exposed side of the wall, the multiplying factor shall be 1.00.

For hollow-core prestressed concrete slabs where the cores are a constant cross section throughout the length, the equivalent thickness is obtained by dividing the net cross-sectional area of the slab (including grout in the joints) by its width. Joints that are between the adjacent precast-concrete slabs are not considered in calculating the slab thickness if the concrete topping is at least 1 inch thick (2.5 cm). If no concrete topping is used, then the joints must be grouted to at least one-third the slab thickness at the joint but not less than 1 inch, or the joints must be made fire-resistant by other approved methods. The slab, reinforced-beam, and prestressed beam covers all must have a minimum thickness that complies with the 2006 International Building Code; please refer to this book for tables that refer to these sections.

The concrete cover for an individual tendon is the minimum thickness of concrete between the surface of the tendon and the fire-exposed surface of the beam, except for ungrouped ducts, where the assumed cover thickness is the minimum thickness of concrete between the surface of the duct and the fire-exposed surface of the beam. The dimensions for reinforced-concrete columns for fire-rated resistance ratings are found in Table 7.9.

Now we are going to touch upon the procedures for calculating the fire-resistance rating of concrete masonry. The equivalent thickness of concrete masonry must be in accordance with this section. The following equation is used in determining the thickness:

$T_e = V_n / LH$

where

T_e = Equivalent thickness of concrete masonry unit (inch) (mm)

V_n = Net volume of masonry unit (inch^3) (mm^2)

L = Specified length of masonry

H = Specified height of masonry

The equivalent thickness, T_e, is the value obtained for the concrete-masonry unit determined in accordance with ASTM C 140. T_e is the solid grouted concrete masonry and the actual thickness. Tables in Chapter 7 of the *International Building Code 2006* contain information and data for the following:

• W/D ratios for steel columns
• Properties of concrete
• Thermal conductivity of concrete or clay
• Weight-to-heated-perimeter ratios
• Fire resistance of concrete-masonry-protected steel columns
• Fire resistance of clay-masonry-protected steel columns
• Minimum cover for steel columns encased in normal-weight concrete
• Minimum cover for steel columns encased in lightweight concrete

TABLE 7.9 Minimum dimension of concrete columns (inches).

TYPES OF CONCRETE	FIRE-RESISTANCE RATING (hours)				
	1	1 1/2	2[a]	3[a]	4[b]
Siliceous	8	9	10	12	14
Carbonate	8	9	10	11	12
Sand-lightweight	8	8 1/2	9	10 1/2	12

• Minimum cover for steel columns in normal weight pre-cast covers

• Minimum cover for steel columns in structural lightweight precast covers

This concludes Chapter 7. Let's move into the next chapter and learn about interior finishes.

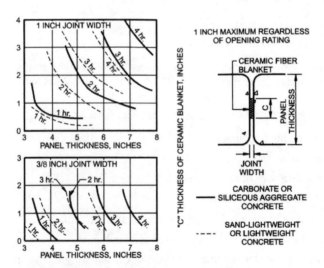

FIGURE 7.3 Ceramic fiber joint protection.

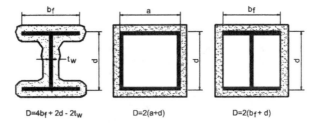

FIGURE 7.9 Determination of the heated perimeter of structural steel columns.

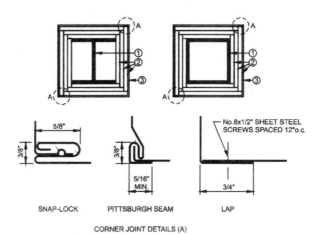

FIGURE 7.10 Gypsum wallboard protected structural steel columns with sheet steel column covers.

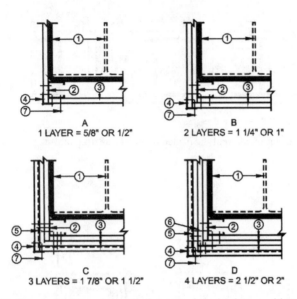

FIGURE 7.11 Gypsum wallboard protected structural steel columns with steel stud/screw attachment system.

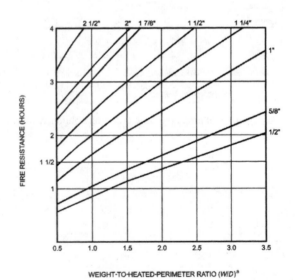

For SI: 1 inch = 25.4 mm, 1 pound per linear foot/inch = 0.059 kg/m/mm.

FIGURE 7.12 Fire resistance of structural steel columns protected with various thicknesses of Type X gypsum wallboard.

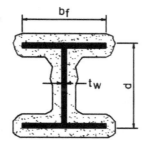

FIGURE 7.13 Wide flange structural steel columns with spray-applied fire-resistant materials.

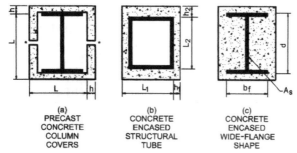

FIGURE 7.14 Concrete protected structural steel columns.

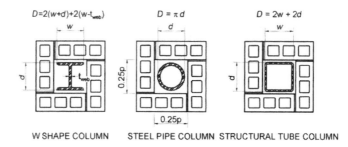

For SI: 1 inch = 25.4 mm.

FIGURE 7.15 Concrete or clay masonry protected structural steel columns.

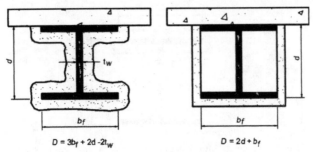

FIGURE 7.16 Determination of the heated perimeter of structural steel beams and girders.

CHAPTER 8
INTERIOR FINISHES

The following requirements limit the allowable flame spread and smoke development based on location and occupancy classification. Exceptions include materials that are less than 0.036 inch (0.9 mm) applied directly to the surface of walls, ceilings, and exposed portions of structural members complying with the requirements for buildings of Type IV construction in Chapter 6. Decorative materials and trim are restricted by combustibility and the flame-spread criteria of NFPA 701 in accordance with Chapter 8 of the code. For buildings that are in flood-hazard areas, as mentioned in Chapter 16, interior finishes, trim, and decorative materials that are below the design flood elevation must be made of flood-damage-resistant materials. All combustible materials may be used as finish for walls, ceilings, floors, and other inside surfaces of buildings. Show windows that are in the outside walls of the first story and are above grade can be made of wood or of unprotected metal framing.

DEFINITIONS

In this section I have added definitions that apply to interior finishes.

• Expanded-Vinyl Wall Covering: Wall covering consisting of a woven textile backing, an expanded vinyl bas- coat layer, and a nonexpanded vinyl skin coat. The expanded base coat is a homogeneous vinyl layer that contains a blowing agent. During processing, the blowing agent decomposes, causing this layer to expand by forming closed cells. The total thickness of the wall covering is approximately 0.055 inch to 0.070 inch (1.4 mm to 1.78 mm).

• Flame Spread: The growth of flame over a surface.

• Flame-Spread Index: A relative measure, expressed as a dimensionless number, made from visual measurements of the spread of flames versus time for a material tested in accordance with ASTM E 84.

!**Code**alert

Show windows in the exterior walls of the first story above grade shall be permitted to be of wood or of unprotected metal framing.

- Interior Floor Finish: The exposed floor surfaces of buildings including coverings applied over a finished floor or stair, including risers.

- Interior Wall and Ceiling Finish: The exposed interior surfaces of buildings, including but not limited to fixed or movable walls and partitions, toilet-room privacy partitions, or other finishes applied structurally or for decoration, acoustical correction, surface insulation, structural fire resistance, or similar purposes, but not including trim.

- Smoke-Developed Index: A comparative measure, expressed as a dimensionless number, made from measurements of smoke obscuration versus time for a material tested in accordance with ASTM E 84.

- Trim: Picture molds, chair rails, baseboards, handrails, door and window frames, and similar decorative or protective materials used in fixed applications.

WALL AND CEILING FINISHES

Walls and ceilings are found in every room in a house, building, or structure. The possibilities of design and decoration are endless. From paint to wallpaper or paneling, there are many ways to personalize your walls. But before you break out the brushes and drop cloth, make sure you know the requirements of the 2006 International Building Code regarding wall and ceiling finishes.

Interior wall and ceiling finishes are classified in accordance with ASTM E 84. Interior finish materials must be grouped in classes according to their flame-spread and smoke-developed indexes. I have provided these for you below:

- Class A: Flame spread 0-25; smoke-developed 0-450

- Class B: Flame spread 26-75; smoke-developed 0-450

- Class C: Flame spread 76-200; smoke-developed 0-450

If your interior walls are made of materials other than textiles, they must be tested in accordance with NFPA 286. During 40 kW exposure of the interior fin-

!Codealert

Foam plastics shall not be used as interior finish or trim except as provided in Section 2603.9 or 2604. This section shall apply both to exposed foam plastics and to foam plastics used in conjunction with a textile or vinyl facing or cover.

ish, the flame cannot spread to the ceiling. During 160 kW exposure of the interior finish, the flame cannot spread to the outer part of the sample on any wall or ceiling, nor can flashover occur. During the entire test exposure of the interior finish, the peak rate of heat release throughout the NFPA 286 test cannot be more than 800 kW. Nor can the total smoke released be higher than 1,000 m². Any interior-finish materials that are regulated by this chapter must be fastened in such a way that the materials cannot become separated if exposed to room temperatures of 200 degrees F (93 degrees C) for no less than 30 minutes.

In cases where interior-finish materials applied on walls, ceilings, or structural parts are required to have a fire-resistance rating or to be made of noncombustible construction, they must comply to the provisions of this section. Anywhere that walls and ceilings are required by the code to be fire-resistance-rated or of noncombustible construction, the interior-finish material has to be directly applied against the construction or furring strips. If you choose to use furring strips, you must be aware that they cannot be more than 1.75 inches (44 mm) long and must be applied directly to the wall surface.

Don't forget that any spaces between furring strips must be filled in with an inorganic or Class A material or be fire-blocked at a maximum of 8 feet (2.4 m) in any direction. If and when walls and ceilings are required to be fire-resistance-

!Codealert

A comparative measure, expressed as a dimensionless number, derived from visual measurements of the spread of flame versus time for a material tested in accordance with ASTM E 84.

!Codealert

A comparative measure, expressed as a dimensionless number, derived from measurements of smoke obscuration versus time for a material tested in accordance with ASTM E 84.

rated or non-combustible construction and if walls are set out or ceilings are dropped more than the distances specified in Chapter 8, Class A finish materials must be used except where they are protected on both sides by an automatic sprinkler system or attached to non-combustible backing or furring strips.

The hangers and assembly members of dropped ceilings that are below the main ceiling line must be made of noncombustible materials, except that in Type III and V construction fire-retardant-treated wood is permitted. Wall and ceiling finishes of all classes that are installed directly against the wood decking, planking, or furring of Type IV construction must be fire-blocked.

Keep in mind that any interior wall or ceiling finish that is no more than 0.25 inch (6.4 mm) thick must be applied directly against a noncombustible backing, except for Class A materials and materials where qualified tests were made with the material suspended or furred out from the noncombustible backing. Table 8.1 contains information regarding interior wall and ceiling finish requirements by occupancy.

INTERIOR FLOOR FINISH

All interior floor finishes and floor-covering materials, with the exception of traditional types such as wood, vinyl, linoleum, or terrazzo and the floor coverings that are made of fibers, must comply with this section of the code. As with interior walls, floor finishes must be tested by an approved agency in accordance with NFPA 253. The floor covering must be identified by a hang tag or other suitable method so that its manufacture is identified.

It is very important that testing and manufacture information is always available. You never know when a building official is going to request to see that information, and he or she is not going to be very happy if you cannot produce it when asked. The building official can use this as a reason to give a stop-work notice until the code officer is satisfied with the appropriate paperwork.

TABLE 8.1 Interior wall and ceiling finish requirements by occupancy[k].

GROUP	SPRINKLERED[l]			NONSPRINKLERED		
	Exit enclosures and exit passageways[a,b]	Corridors	Rooms and enclosed spaces[c]	Exit enclosures and exit passageways[a,b]	Corridors	Rooms and enclosed spaces[c]
A-1 & A-2	B	B	C	A	A[d]	B[e]
A-3[f], A-4, A-5	B	B	C	A	A[d]	C
B, E, M, R-1, R-4	B	C	C	A	B	C
F	C	C	C	B	C	C
H	B	B	C[g]	A	A	B
I-1	B	C	C	A	B	B
I-2	B	B	B[h,i]	A	A	B
I-3	A	A[j]	C	A	A	B
I-4	B	B	B[h,i]	A	A	B
R-2	C	C	C	B	B	C
R-3	C	C	C	C	C	C
S	C	C	C	B	B	C
U	No restrictions			No restrictions		

For SI: 1 inch = 25.4 mm, 1 square foot = 0.0929 m².

a. Class C interior finish materials shall be permitted for wainscotting or paneling of not more than 1,000 square feet of applied surface area in the grade lobby where applied directly to a noncombustible base or over furring strips applied to a noncombustible base and fireblocked as required by Section 803.4.1.

b. In exit enclosures of buildings less than three stories in height of other than Group I-3, Class B interior finish for nonsprinklered buildings and Class C interior finish for sprinklered buildings shall be permitted.

c. Requirements for rooms and enclosed spaces shall be based upon spaces enclosed by partitions. Where a fire-resistance rating is required for structural elements, the enclosing partitions shall extend from the floor to the ceiling. Partitions that do not comply with this shall be considered enclosing spaces and the rooms or spaces on both sides shall be considered one. In determining the applicable requirements for rooms and enclosed spaces, the specific occupancy thereof shall be the governing factor regardless of the group classification of the building or structure.

d. Lobby areas in Group A-1, A-2 and A-3 occupancies shall not be less than Class B materials.

e. Class C interior finish materials shall be permitted in places of assembly with an occupant load of 300 persons or less.

f. For places of religious worship, wood used for ornamental purposes, trusses, paneling or chancel furnishing shall be permitted.

g. Class B material is required where the building exceeds two stories.

h. Class C interior finish materials shall be permitted in administrative spaces.

i. Class C interior finish materials shall be permitted in rooms with a capacity of four persons or less.

j. Class B materials shall be permitted as wainscotting extending not more than 48 inches above the finished floor in corridors.

k. Finish materials as provided for in other sections of this code.

l. Applies when the exit enclosures, exit passageways, corridors or rooms and enclosed spaces are protected by a sprinkler system installed in accordance with Section 903.3.1.1 or 903.3.1.2.

!Codealert

See Section 803.6.2 for many changes in the code.

COMBUSTIBLE MATERIALS

This section of the code is about combustible materials installed on or embedded in floors of buildings that are of Type I or II construction, with the exception of stages and platforms, which have to comply with Chapter 4. The subfloor construction that the regulations apply to includes floor sleepers, bucks, and nailing blocks. They do not need to be made of noncombustible materials unless the space between the fire-resistance-rated floor and the flooring is either filled with approved noncombustible materials or fire-blocked. The code allows wood floor finishes to be attached directly to the embedded or fire-blocked wood sleepers, and you may also cement them directly to the top surface of the fire-resistance-rated floor or even to a wood subfloor attached to sleepers. If you are using combustible insulating boards that are not more than 1/2 inch (12.7 mm) thick, approved finish flooring is allowed if directly attached to a noncombustible floor.

DECORATIVE MATERIALS AND TRIM

This section of the code applies to items such as curtains, draperies, hangings, and other materials used for decoration. Occupancies in Groups A, E, and R-1 and dormitories in Group R-2 must comply with this section. Any decorative materials, such as those mentioned above, must meet the flame-spread performance criteria of NFPA 701 or be made of noncombustible materials. This applies to Groups I-1 and I-2 unless the decorative materials include but are not limited to photographs and paintings. In Group I-3, combustible decorative materials are not allowed at all. In Groups B and M, fabric partitions that are hung from ceilings but not supported by the floor must meet the flame-spread performance requirements of NFPA 701 or be noncombustible.

Foam plastics that are used as trim in any occupancy must comply with Chapter 26. Imitation leather or other material that consists of is or coated with a hazardous base must not be used in Group A occupancies. Lastly, combustible trim, except for handrails and guardrails, cannot comprise more than 10 percent of the total wall or ceiling area where it is located.

CHAPTER 9
FIRE-PROTECTION SYSTEMS

This chapter specifies where fire-protection systems are required and how they must be designed, installed, and operated. I have included definitions exclusive to this chapter's subject. When installing, repairing, operating, and maintaining fire-protection systems, your work must be in accordance with the International Building Code and the International Fire Code. Even modifications are required to follow these codes; you must also get approval from the building official—there are no exceptions to this rule.

If you are using or installing a fire-protection system or part of one that is not required by this code, it also must meet code requirements. When you are installing threads for connections to sprinkler systems, standpipes, yard hydrants, or any other fire-hose connections, you must make certain that they are compatible with the connections that are used by the local fire department. A simple phone call or visit can save time and money or even avoid a code decision against you by ensuring that these connections are a match.

All fire-protection systems must be tested in accordance with the requirements of the IBC and the International Fire Code. If testing is required, it is your responsibility that it is conducted in the presence of the code official. It is not enough to have the test(s) done; the building official must observe the tests from start to finish. The costs of these tests are the responsibility of the owner or a person representing the owner. It is very important that you realize that it is illegal to occupy a structure or even a portion of a structure if the proper testing has not been done and/or approved by the building official.

Fire-protection systems must be monitored (where required) by an approved supervising station in accordance with NFPA 72. All automatic sprinkler systems must also be monitored by an approved supervising station, with the exception of systems protecting one- and two-family housing units and limited-area systems that serve fewer than 20 sprinklers.

!Definitionalert

Alarm-Notification Appliance: A fire-alarm system component such as a bell, horn, speaker, light, or text display that provides audible, tactile, or visible outputs or any combination of these.

Alarm-Verification Feature: A feature of automatic fire-detection and alarm systems to reduce unwanted alarms in which smoke detectors report alarm conditions for a minimum period of time or confirm alarm conditions within a given time period after being automatically reset in order to be accepted as a valid alarm-initiation signal.

Audible Alarm-Notification Appliance: A notification appliance that alerts through the sense of hearing.

Fire-alarm systems require a similar monitoring by an approved supervising station with the exception of single- and multiple-station smoke alarms and smoke detectors in Group I-3 occupancies. Manual fire alarms, automatic fire-extinguishing systems, and emergency alarm systems in Group H occupancies must be monitored by an approved supervising station as well.

When approved by the building official, on-site monitoring at a constantly attended location may be permitted. This is true only if the owner agrees that notifications to the fire department will be equal to those provided by an approved supervising station.

You will find definitions that pertain to this topic scattered throughout this chapter.

AUTOMATIC SPRINKLER SYSTEMS

All automatic sprinkler systems installed in a structure must follow the rules of this section. If you plan on or are using alternative automatic fire-extinguishing systems complying with this chapter of the code, they must be recognized by the applicable standard and approved by the fire-code official. All new buildings and structures must be provided with approved automatic sprinkler systems that comply with the code. In this chapter I cover classifications of buildings by group.

Group A

Group A-1 occupancies require an automatic sprinkler system where one or more of the following conditions exist:

• The fire area exceeds more than 12,000 square feet (1,115 m²).

• The fire area has an occupant load of more than 300 people.

• The fire area is located on a floor other than the level of exit doors.

• The fire area contains a multitheater complex.

The following list pertains to Group A-2 occupancies and the conditions in which automatic sprinkler systems are mandatory:

• The fire area exceeds 5,000 square feet (465 m²).

• The fire area has an occupant load of 100 or more.

• The fire area is located on a floor other than the level of exit doors.

An automatic sprinkler system must be provided for Group A-3 and Group A-4 occupancies when the following conditions exist:

• The fire area exceeds 12,000 square feet.

• The fire area has an occupant load of 300 or more.

• The fire area is located on a floor other that the level of exit doors.

Group A-3 and A-4 have the same exception to these rules: any areas used exclusively as participant sport areas where the main floor area is located at the same level as the level of the main entrance and exit.

Group A-5 occupancies must be equipped with an automatic sprinkler system in the following areas:

• Concession stands

• Retail areas

• Press boxes

• Other accessory-use areas in excess of 1,000 square feet (93 m²)

Group E

Group E occupancies require automatic sprinklers throughout all areas that are greater than 20,000 square feet (1858 m²) and throughout every portion below the level of exit discharge of buildings used for educational purposes. The exception is that an automatic sprinkler system is not required in any fire area or any area below the level of exit discharge when every classroom throughout the building has at least one exterior exit door at ground level.

Group F

In Group F-1 occupancies automatic sprinkler systems must be provided throughout all buildings when any of the following conditions exists:

• The fire area is more than 12,000 square feet (1,115 m²).

• The fire area is located more than three stories above grade plane.

• The combined area of all fire areas on all floors, including any mezzanines, is more than 24,000 square feet (2,230 m^2).

Group F-1 occupancies such as those containing woodworking operations in excess of 2,500 square feet (232 m^2) in areas that yield finely divided combustible waste or use finely divided combustible materials are required to have automatic sprinkler systems installed.

Group H

High-hazard occupancies, such as Group H, must have an automatic sprinkler installed. The design of the sprinkler system cannot be less than that required for the occupancy hazard classification in accordance with Table 9.1. Where the design area of the sprinkler system consists of a corridor protected by one row of sprinklers, the maximum number of sprinklers required is 13. Keep in mind that buildings in which cellulose-nitrate film or pyroxylin plastics are manufactured, stored, or handled in quantities of more than 100 pounds are required to be equipped with automatic sprinklers.

Group I

Institutional buildings, which are classified as Group I, are required to install automatic sprinkler systems. Any automatic sprinkler systems that meet NFPA 13R requirements and are installed in accordance to code are allowed to be installed in Group I.

Group M

Group M occupancies that meet one of the following conditions are required to have an automatic sprinkler system installed throughout the building:

• Any fire area that is more than 12,000 square feet (1,115 m^2)

• A fire area that is located more than three stories above grade plane

• A combined fire area on all floors, including any mezzanines, that exceeds 24,000 square feet (2,230 m^2)

!Definitionalert

Constant Attended Location: A location designated at a facility that is staffed on a continuous basis by trained personnel. Alarms and supervisory signals are monitored, and facilities are provided for notification of the fire department or other emergency services.

TABLE 9.1 Group H-5 sprinkler design criteria.

LOCATION	OCCUPANCY HAZARD CLASSIFICATION
Fabrication areas	Ordinary Hazard Group 2
Service corridors	Ordinary Hazard Group 2
Storage rooms without dispensing	Ordinary Hazard Group 2
Storage rooms with dispensing	Extra Hazard Group 2
Corridors	Ordinary Hazard Group 2

Buildings of Group M occupancies in which storage of merchandise is piled high or put into rack-storage arrays are required to have an automatic sprinkler system that must be in accordance with the International Fire Code.

Group R

All Group R buildings are required to provide automatic sprinklers throughout.

Group S

In Group S-1 the entire building is required to be equipped with an automatic sprinkler system where one of the following conditions exists:

• The fire area is more than 12,000 square feet (1,115 m²)

• A fire area is located more than three stories above grade plane.

• The combined fire area on all floors, including any mezzanines, is more than 24,000 square feet (2,230 m²).

Repair garages that are two or more stories in height, including basements, more than 10,000 square feet in area, more than 12,000 square feet in area within a one-story building, or within a basement are required to have automatic sprinkler systems installed. This not only protects your customers and their personal belongings but yours as well. If you are storing tires at a bulk rate and the area of storage is over 20,000 cubic feet, it is absolutely necessary that an automatic sprinkler be installed. There is no exception.

Group S-2 buildings that are classified as enclosed parking garages or that are located beneath other groups, with the exception of enclosed parking garages that are located beneath Group R-3 occupancies, must be furnished with an automatic sprinkler system. This is also true for commercial parking garages and buildings used for storage of commercial trucks or buses where the fire area is more than 5,000 square feet.

Occupancies that have windowless stories always require automatic sprinkler systems. Basements without openings are classified as a windowless story. If your basement has a floor area of more than 1,500 feet and there is not at least one of the following types of exterior wall opening, you must install an automatic sprinkler system:

• Openings below grade must lead directly to ground level by an exterior stairway or an outside ramp that complies with Chapter 10 of the code.

• Openings must be located in each 50 linear feet, or fraction, of the outside wall in the story on at least one side.

• Openings that are entirely above the adjoining ground level must total 20 square feet in each 50 linear feet of exterior wall on at least one side.

These openings and access areas must have a minimum dimension of no less than 30 inches. The openings must also be accessible to the fire department from the outside and must not be obstructed in such a manner that firefighting or rescue cannot be accomplished from the outside. If openings in a story are provided on only one side and the opposite wall is more than 75 feet from those openings, the story must be equipped throughout with an approved automatic sprinkler system, or openings as specified above must be provided on at least two sides of the story.

If any portion of a basement is located more than 75 feet from openings required by this code chapter, the basement must be equipped throughout with an automatic sprinkler system. Although not technically a windowless floor, rubbish and laundry or linen chutes and the rooms in which they terminate require an automatic sprinkler system. Any chute that goes through three or more floors must have additional sprinkler heads installed at alternate floors of the chutes. While in-

!Definitionalert

Automatic Sprinkler System: A sprinkler system, for fire-protection purposes, is an integrated system of underground and overhead piping designed in accordance with fire-protection engineering standards. The system includes a suitable water supply. The portion of the system above the ground is a network of specially sized, hydraulically designed piping installed in a structure or area, generally overhead, to which automatic sprinklers are connected in a systematic pattern. The system is usually activated by heat from a fire and discharges water over the fire area.

!Definitionalert

Multiple-Station Alarm Device: Two or more single-station alarm devices that are capable of interconnection such that actuation of one causes all integral or separate audible alarms to operate. It also can consist of one single-station alarm device with connections to other detectors or to a manual fire-alarm box.

stalling these additional heads it would be wise to keep in mind that they need to be accessible for servicing.

Buildings that are 55 feet or more in height, with the exception of airport-control towers, open parking structures, or occupancies in Group F-2; and have an occupant load of 30 or more; and are located 55 feet or more above the lowest level of fire-department-vehicle access always require an automatic sprinkler system to be installed.

Any automatic sprinkler systems that are required during construction, alteration, and demolition operations must be provided for in accordance with the International Fire Code. Provisions of the IBC that require that a building or portion be equipped throughout with an automatic sprinkler system in accordance with this chapter must also be installed in accordance with NFPA 13 unless otherwise specified.

Automatic sprinklers are not required in the following rooms or areas if they are protected with an automatic fire-detection system, in accordance with this chapter, that will respond to visible or invisible particles of combustion:

- Any room where the application of water or flame and water constitutes a serious life or fire hazard
- Any room or space where sprinklers are considered undesirable because of the nature of the contents, when approved by the fire-code official
- Generator and transformer rooms separated from the remainder of the building by walls and floor/ceiling or roof/ceiling assemblies having a fire-resistance rating of not less than two hours
- Rooms or areas that are of noncombustible construction with entirely noncombustible contents

Sprinklers may be omitted from any room simply because it is damp, is of fire-resistance-rated construction, or contains electrical equipment.

When allowed in buildings of Group R, up to and including four stories in height, automatic sprinkler systems must be installed throughout in accordance with NFPA 13R.

Type V Construction

Sprinkler protection must be provided for exterior balconies, decks, and ground-floor patios of dwelling units where the building is of Type V construction. Sidewall sprinklers that are used to protect such areas must be located so that their deflectors are within 1 to 6 inches below the structural members and a maximum distance of 14 inches below the deck of exterior balconies and decks constructed of open wood joists. Where allowed, automatic sprinkler systems in one-and two-family dwellings must be installed throughout in accordance with NFPA 13 D.

MONITORING AND ALARMS

It is also important to look at sprinkler-system monitoring and alarms. All valves controlling the water supply for automatic sprinkler systems, pumps, tanks, water levels and temperatures, critical air pressures, and water-flow switches on all sprinkler systems must be electrically supervised. Below is a list of exceptions to this rule:

• Automatic sprinkler systems protecting one-and two-family dwellings

• Limited-area systems serving fewer than 20 sprinklers

• Automatic sprinkler systems installed in accordance with NFPA 13R where a common supply main is used to supply both domestic water and the automatic sprinkler systems and a separate shutoff valve for the automatic sprinkler system is not provided

• Jockey pump-control valves that are sealed or locked in the open position

• Control valves to commercial kitchen hoods, paint spray booths, or dip tanks that are sealed or locked in the open position

!Codealert

When approved by the fire-code official, mechanical smoke control for large enclosed volumes, such as atriums or malls, shall be permitted to utilize the exhaust method. Smoke-control systems using the exhaust method shall be designed in accordance with NFPA 92B.

• Valves controlling the fuel supply to fire pump engines that are sealed or locked in the open position

• Trim valves to pressure switches in dry, preaction, and deluge sprinkler systems that are sealed or locked in the open position

Alarm, supervisory, and trouble signals must be distinctly different and automatically sent to an approved central station, remote supervising station, or proprietary supervising station as defined in NFPA 72 or, when approved by the fire-code official, must sound an audible signal at a constantly attended location. There are two exceptions to this rule. The first exception states that underground key or hub valves in roadway boxes provided by the municipality or public utility are not required to be monitored. The second exception states that backflow-prevention-device test valves located in limited-area sprinkler-system supply piping must be locked in the open position. In occupancies required to be equipped with a fire-alarm system, the backflow-preventer valves must be electrically supervised by a tamper switch installed in accordance with NFPA 72 and separately annunciated.

Alarms, such as audible devices, must be approved and connected to every automatic sprinkler system. These sprinkler water-flow alarm devices must be activated by water flow equivalent to the flow of a single sprinkler of the smallest orifice size installed in the system.

Alarm devices must be provided on the outside of the building in an approved location. Where a fire-alarm system is installed, the motion of the automatic sprinkler system must set off the building fire-alarm system. Don't forget this very important detail: all sprinkler systems must be tested and maintained in accordance with the International Fire Code.

ALTERNATIVE SYSTEMS

Any automatic fire-extinguishing systems, other than automatic sprinkler systems, must be designed, installed, inspected, tested, and maintained in accordance with this section and the applicable standards. Any automatic fire-extinguishing systems installed as an alternative to the required automatic sprinkler systems of this chapter must be approved by the fire-code official. You may not use automatic fire-extinguishing systems as alternatives for the purpose of exceptions or reductions that are allowed by other requirements of this code. What this means is that a building owner cannot substitute an alternate system in the hope of applying it to an exception to the code.

The code is written to protect the property and occupancies. Remember that the building-code official can withdraw any permits if is believed that the code has been manipulated in any way in your favor. Before conducting final acceptance tests, the following items must be inspected:

• Hazard specification for consistency with design hazard

• Type, location, and spacing of automatic- and manual-initiating devices

• Size, placement, and position of nozzles or discharge holes

• Location and identification of audible and visible alarm devices

• Identification of devices with proper designations

• Operating instructions

All notification appliances, connections to fire-alarm systems, and connections to approved supervising stations must be tested in accordance with this chapter of the code to verify proper operation. Also keep in mind that the audibility and visibility of notification appliances signaling agent discharge or system operation, where required, must be verified. These, too, need to be tested to verify proper identification and re-transmission of alarms from automatic fire-extinguishing systems.

If you plan to use wet- or dry-chemical extinguishing systems, they must be installed, maintained, periodically inspected, and tested in accordance with NFPA 17. Foam-extinguishing systems must be installed, maintained, periodically inspected, and tested in accordance with NFPA 11 and NFPA 16.

COMMERCIAL COOKING

Occupancies that have commercial-cooking systems must have fire protection and exhaust of a type approved for this use. Preengineered automatic dry- and wet-chemical extinguishing systems must be tested in accordance with UL 300 and listed and labeled for the intended application.

Automatic fire-extinguishing systems of the following types must be installed in accordance with the referenced standard indicated, as follows:

• Carbon-dioxide extinguishing systems, NFPA 12

• Automatic sprinkler systems, NFPA 13

• Foam-water sprinkler or spray systems, NFPA 16

• Dry-chemical extinguishing systems, NFPA 17

• Wet-chemical extinguishing systems, NFPA 17A

!Definitionalert

Standpipe System: An arrangement of piping valves, hose connections, and allied equipment installed in a building or structure with the hose connections located in such a manner that water can be discharged in streams or spray patterns through attached hoses and nozzles.

An exception to this code is factory-built commercial-cooking recirculating systems that are tested in accordance with UL 710B and listed, labeled, and installed in accordance with Section 304.1 of the International Mechanical Code.

Commercial cooking systems must have a manual actuation device located at or near a means of exit from the cooking area and a minimum of 10 feet and a maximum of 20 feet from the kitchen exhaust system. The manual actuation device must be installed not more than 48 or less than 42 inches above the floor and must clearly identify the hazard protected. The manual motion must require a maximum force of 40 pounds and a maximum movement of 14 inches to set the fire-suppression system in motion.

Automatic sprinkler systems are not required to be equipped with manual movement means. The fire-suppression system must automatically shut down the fuel or electrical power supply to the cooking equipment, and the supply must be reset manually. If carbon-dioxide systems are used, there must be a nozzle at the top of the ventilating duct. Additional nozzles that are balanced to give uniform distribution must be installed within vertical ducts more than 20 feet and horizontal ducts more than 50 feet.

Dampers must be installed at either the top or the bottom of the duct and must be arranged to operate automatically upon activation of the fire-extinguishing system. If the damper is installed at the top of the duct, the top nozzle must be immediately below the damper.

The sizing of an automatic carbon-dioxide fire-extinguishing system is extremely important to protect against all hazards venting through a common duct at the same time. If your cooking equipment is of a commercial type, it must be arranged to shut off the ventilation system upon activation.

STANDPIPE SYSTEMS

Standpipe systems must be provided in new buildings and structures in accordance with this chapter of the code. Fire-hose threads used in connection with standpipe systems must be approved and must be compatible with fire-department hose threads. The location of fire-department hose connections must be approved as well.

All standpipe systems must be installed in accordance with this section of the code and NFPA 14. Standpipe systems are allowed to be combined with automatic sprinkler systems but are not required in Group R-3 occupancies.

Class I

Class I standpipes have many conditions associated with their use:

• Class I standpipes are allowed in buildings equipped throughout with an automatic sprinkler system in accordance with this chapter.

- Class I manual standpipes are allowed in open parking garages where the highest floor is located not more than 150 feet above the lowest level of fire-department-vehicle access.

- Class I manual dry standpipes are allowed in open parking garages that are subject to freezing temperatures, provided that the hose connections are located as required for Class II standpipes in accordance with this section.

- Class I standpipes are allowed in basements equipped throughout with an automatic sprinkler system.

- In determining the lowest level of fire-department-vehicle access, you need not consider recessed loading docks for four vehicles or less or conditions where topography makes access from the fire-department vehicle to the building impractical or impossible.

Class I automatic wet standpipes must be provided in nonsprinklered Group A buildings with an occupant load exceeding 1,000 persons except for open-air-seating spaces that are without enclosed spaces. Class I automatic dry, semi-automatic dry, or manual wet standpipes are allowed in buildings where the highest floor surface used for human occupancy is 75 feet or less above the lowest level of fire-department-vehicle access.

Some groups have special requirements. The code clearly states that a building such as a covered mall, for example, must be equipped throughout with a standpipe system where required by this chapter. As the owner or contractor of a building you must provide the following hose connections in the locations required:

- Within the mall at the entrance to each exit passageway or corridor

- At each floor-level landing within enclosed stairways opening directly on the mall

- At exterior public entrances to the mall

Hose connections must be equipped with an approved adjustable fog nozzle and be mounted in a cabinet or on a rack. Make yourself aware that all underground buildings must be equipped throughout with a Class I automatic or manual wet standpipe system.

Buildings that are equipped with a helistop or heliport that contains a standpipe must extend the standpipe to the roof level on which the helistop or heliport is located in accordance with Section 1107.5 of the International Fire Code. Marinas and boatyards must also be equipped throughout with standpipe systems in accordance with NFPA 303. The following list contains all locations that Class I standpipe hose connections must be provided:

- In every required stairway, a hose connection must be provided for each floor level above or below grade.

- Hose connections must be located at an intermediate floor-level landing between floors unless otherwise approved by the fire-code official.

• Hose connections must be located on each side of the wall adjacent to the exit opening of a horizontal exit, except where floor areas that are adjacent to a horizontal exit are reachable from the exit stairway. Hoses must yield a 30-foot stream from a nozzle attached to 100 feet of hose. A hose connection must not be required at the horizontal exit.

• At the entrance from an exit passageway to other areas of a building a hose connection is required.

• In covered mall buildings, connections must be installed adjacent to each exterior public entrance to the mall and adjacent to each entrance from an exit passageway or exit corridor to the mall.

• Where the roof has a slope less than four units vertical in 12 units horizontal, each standpipe must be provided with a hose connection located either on the roof or at the highest landing of stairways with access to the roof. An additional hose connection must be provided at the top of the most hydraulically remote standpipe for testing purposes.

• Where the most remote portion of a nonsprinklered floor or story is more than 200 feet form a hose connection, the fire-code official is authorized to require that additional hose connections be provided in approved locations.

Class II and Class III

Class II standpipe hose connections must be accessible and located so that all portions of the building are within 30 feet of a nozzle attached to 100 feet of hose. Groups A-1 and A-2 occupancies with occupant loads of more than 1,000 require hose connections located on each side of any stage, on each side of the rear of the auditorium, on each side of the balcony, and on each tier of dressing rooms. Fire-resistance-rated protection of riser and laterals of Class II standpipe systems is not required, but for Class II systems a minimum 1-inch hose may be used for hose stations in light-hazard occupancies where investigated and listed for this service and where approved by the fire-code official.

Any stage that is greater than 1,000 square feet in area must be equipped with a Class III wet-standpipe system with 1.5-inch and 2.5-inch hose connections on each side of the stage. The exception to this is where the building or area is equipped throughout with an automatic sprinkler system. The 1.5-inch hose connection must be installed in accordance with NFPA 13 or in accordance with NFPA 14 for Class II or III standpipes. The 1.5-inch hose connections must be equipped with sufficient lengths of hose to provide fire protection for the stage area.

Class III standpipe systems must be installed throughout buildings where the floor level of the highest story is located more than 30 feet above the lowest level of fire-department-vehicle access or where the floor level of the lowest story is located more than 30 feet below the highest level of fire-department-vehicle access. The location of Class III standpipe hose connections are the same requirements as those for Class I standpipes and must have Class II hose connections as required in this chapter. You must not forget that risers and laterals of Class III standpipe systems must be protected as required for Class I systems as well. In building where more than one Class III standpipe is provided, the standpipes must be interconnected at the bottom.

As with all code regulations, you must contact your local building-code official for any additional information or with any questions you may have. The IBC is designed to protect everyone, so please be sure that you understand the code and question it when needed.

PORTABLE FIRE EXTINGUISHERS

All occupancies and locations must be equipped with a portable fire extinguisher as required by the International Fire Code.

ALARM AND DETECTION SYSTEMS

This section covers the application, installation, performance, and maintenance of fire-alarm systems and their components. Construction documents for fire-alarm systems must be presented for review and approval before you install any system. I don't believe that you would want to install anything and then find out that it does not meet code. You, as a building owner or contractor, stand to lose a lot of time, effort, and money if you do not follow the code as written. Construction documents must include but are not be limited to all of the following:

• A floor plan that indicates the use of all rooms

• Locations of alarm-initiating and -notification appliances

• Alarm-control and trouble-signaling equipment

• Annunciation

• Power connection

• Battery calculation

• Conductor type and sizes

• Voltage-drop calculation

• Manufacturers, model numbers, and listing information for equipment, devices, and materials

• Details of ceiling height and construction

• The interface of fire-safety control functions

Systems and their components must be listed and approved for the purpose for which they are installed. An approved manual, automatic, or manual and automatic fire-alarm system installed in accordance with the provisions of the code and NFPA 72 must be provided in new buildings and structures in accordance with this chapter, where applicable, and provide occupant notification unless other requirements are provided by another chapter or section of this code.

Automatic-sprinkler protection that is installed in accordance with this section is provided and connected to the building fire-alarm system; automatic heat detection is not required in this instance. The automatic fire detectors must be smoke detectors. Where ambient conditions prohibit installation of automatic smoke detection, other automatic fire detection must be allowed.

Group A

Group A occupancies with an occupant load of 300 people or more must have a manual fire-alarm system in play. Portions of Group A occupancies used for assembly purposes must be provided with a fire-alarm system as required for that group. However, take note that manual fire-alarm boxes are not required where the building is equipped throughout with an automatic sprinkler system and the alarm-notification appliance will activate upon water flow from the sprinkler system.

Activation of the fire alarm in Group A occupancies with an occupant load of 1,000 people or more must initiate a signal using an emergency voice or alarm-communication system in accordance with NFPA 72, unless, where approved, the prerecorded announcement is allowed to be manually deactivated for a period of time but not more than three minutes for the sole purpose of allowing a live voice announcement from an approved and constantly attended location. Any emergency voice or alarm communication systems must be provided with an approved emergency power source.

Group B

Group B occupancies that have an occupant load of more than 100 persons above or below the lowest level of exit discharge are required to install a manual fire-alarm system except where the building is equipped throughout with an automatic sprinkler system and the alarm-notification appliances will activate upon sprinkler water flow.

!Codealert

The height of the lowest horizontal surface of the accumulating smoke layer shall be maintained at least 7 feet above any walking surface that forms a portion of a required egress system within the smoke zone.

Group E

Group E occupancies must also have a manual fire-alarm system; when automatic sprinkler systems or smoke detectors are installed, these systems must be connected to the building fire-alarm systems. The following list contains the exceptions to the above:

• Group E occupancies that have an occupant load of less than 50 people

• Interior corridors protected by smoke detectors with alarm verification

• Auditoriums, cafeterias, gymnasiums, and similar occupancies protected by heat detectors or other approved devices

• Off-premises monitoring provided

• The capability to activate the evacuation signal from a central point provided

• In buildings where normally occupied spaces are provided with a two-way communication system between such spaces and a constantly attended receiving station from where a general evacuation alarm can be sounded, except in locations specifically designated by the fire-code official

Manual fire-alarm boxes are not required in Group E occupancies where the building is equipped throughout with an approved automatic sprinkler system, the notification appliances will activate on sprinkler water flow, and manual activation is provided from a normally occupied location.

Group F

Group F occupancies that are two or more stories in height and have an occupant load of 500 people or more above or below the lowest level of exits must have a manual fire-alarm system. Manual fire-alarm boxes are not required if an automatic sprinkler system is installed and if the system activates upon sprinkler water flow.

Group H

Group H occupancies, which contain highly toxic gases, organic peroxides, and oxidizers, are required by the code, in accordance with Chapters 37, 39, and 40 of the International Fire Code, to install a manual fire-alarm system.

Group I

Group I occupancies are required to have a manual fire-alarm system. However, this does not pertain to resident or patient sleeping areas of Group I-1 and I-2 occupancies, provided that all nurses' control stations or other constantly attended staff locations are visible and continuously accessible and that travel distances that are required by the code are not exceeded. Corridors, habitable spaces other than sleeping units and kitchens, and waiting areas that are open to corridors must be

control systems. It is importantto be able to control smoke during a fire incident so that the danger is contained and the building can be safely evacuated. This section applies to mechanical or passive smoke-control systems when they are required by other provisions of the code. The purpose of this section is to establish the minimum requirements for the design, installation, and testing of smoke-control systems that are intended to provide a tenable environment for the evacuation or relocation of occupants.

Don't misinterpret this code as relating to assistance in fire suppression or overhaul activities. Smoke-control systems that are regulated by this section serve a different purpose than the smoke- and heat-venting provisions that will be discussed in the next section. Also keep in mind that mechanical smoke-control systems must not be considered exhaust systems under Chapter 5 of the International Mechanical Code.

Buildings, structures, or parts thereof are required to have a smoke-control system designed in accordance with the applicable requirements of this section and the generally accepted and well-established principles of engineering relevant to the design. All construction documents must include sufficient information and detail to adequately describe the elements of the design necessary for the proper implementation of the smoke-control systems. These documents must be accompanied by sufficient information and analysis to demonstrate compliance with these provisions.

All buildings and structures are required to have ordinary inspections and test requirements, but smoke-control systems are mandated to undergo special inspections and tests sufficient to verify the proper commissioning of the design in its final installed condition. You must include detailed procedures and methods to be used and the items subject to these inspections and tests on the design submission documents. This commissioning must be in accordance with generally accepted engineering practice and, where possible, based on published standards for the particular testing that is involved.

The smoke-control system must be supplied with two sources of power. Primary power must be from the normal building power system. Secondary power must be from an approved standby source complying with the ICC Electrical Code.

The standby power source and its transfer switches must be in a separate room from the normal power transformers and switch gear. It must be enclosed in a room that is constructed of not less than 1-hour fire barriers and is ventilated directly to and from the outside. All power distribution from the two sources must come from independent routes. The transfer to full standby power has to be automatic and within 60 seconds of failure of the primary power. This system is required to comply with the IBC or the ICC Electrical Code. It is imperative that mechanical smoke-control systems include provisions for verification. A verification list follows:

- Positive confirmation of actuation
- Testing
- Manual override
- The presence of power downstream of all disconnects
- A preprogrammed weekly test sequence
- Report abnormal conditions audibly, visually, and by printed report

Activation of smoke-control systems must follow code regulations. Mechanical smoke-control systems that use pressurization, airflow, or exhaust methods must have completely automatic control. Passive smoke-control systems actuated by approved spot-type detectors listed for releasing service are permitted. If your smoke-control system is completely automatic, the automatic-control sequence has to be initiated from an appropriately zoned automatic sprinkler system complying with this section, manual controls that are readily accessible to the fire department, and any smoke detectors required by engineering analysis.

Don't forget that any detection and control systems must be clearly marked at all junctions, accesses, and terminations, and identical control diagrams showing all devices in the system and identifying their location and function must be maintained and kept on file with the fire-code official, the fire department, and the fire command center in a format and manner that has been approved by the fire chief.

A firefighter's smoke-control panel, for fire-department emergency-response purposes only, must be provided and has to include manual control or override of automatic control for mechanical smoke-control systems. The panel will be located in a fire command center in high-rise buildings or buildings with smoke-protected assembly seating. In all other buildings, the firefighter's smoke-control panel must be installed in an approved location adjacent to the fire-alarm control panel. The firefighter's smoke- control panel must comply with Chapter 9 of the 2006 International Building Code.

With all these fire- and smoke-alarm systems in place, we cannot forget smoke and heat vents. Where smoke and heat vents, exhaust systems, or draft curtains are required by the code or otherwise installed, they must conform to the requirements of the section. There are two exceptions to this. Frozen-food warehouses used solely for storage of Class I and II merchandise, where protected by an approved automatic sprinkler system, or areas of buildings that are equipped with early-suppression fast-response (ESFR) sprinklers do not require automatic smoke and heat vents.

Smoke and heat vents must be installed in the roofs of Group F-1 or S-1 buildings that have occupancies or that are more than 50,000 square feet, except for Group S-1 aircraft repair hangars. The design and installation of smoke and heat vents and draft curtains can be seen in Table 9.3.

TABLE 9.3 Requirements for draft curtains and smoke and heat vents[a].

OCCUPANCY GROUP AND COMMODITY CLASSIFICATION	DESIGNATED STORAGE HEIGHT (feet)	MINIMUM DRAFT CURTAIN DEPTH (feet)	MAXIMUM AREA FORMED BY DRAFT CURTAINS (square feet)	VENT-AREA-TO-FLOOR-AREA RATIO[c]	MAXIMUM SPACING OF VENT CENTERS (feet)	MAXIMUM DISTANCE TO VENTS FROM WALL OR DRAFT CURTAINS[b] (feet)
Group F-1 and S-1	—	$0.2 \times H^d$ but ≥ 4	50,000	1:100	120	60
High-piled Storage (see Section 910.2.3) I-IV (Option 1)	≤ 20	6	10,000	1:100	100	60
	> 20 ≤ 40	6	8,000	1:75	100	55
High-piled Storage (see Section 910.2.3) I-IV (Option 2)	≤ 20	4	3,000	1:75	100	55
	> 20 ≤ 40	4	3,000	1:50	100	50
High-piled Storage (see Section 910.2.3) High hazard (Option 1)	≤ 20	6	6,000	1:50	100	50
	> 20 ≤ 30	6	6,000	1:40	90	45
High-piled Storage (see Section 910.2.3) High hazard (Option 2)	≤ 20	4	4,000	1:50	100	50
	> 20 ≤ 30	4	2,000	1:30	75	40

For SI: 1 foot = 304.8 mm, 1 square foot = 0.0929 m².

a. Requirements for rack storage heights in excess of those indicated shall be in accordance with Chapter 23 of the International Fire Code. For solid-piled storage heights in excess of those indicated, an approved engineered design shall be used.

b. The distance specified is the maximum distance from any vent in a particular draft curtained area to walls or draft curtains which form the perimeter of the draft curtained area.

c. Where draft curtains are not required, the vent-area-to-floor-area ratio shall be calculated based on a minimum draft curtain depth of 6 feet (Option 1).

d. "H" is the height of the vent, in feet, above the floor.

FIRE COMMAND CENTER

I want to talk about the code pertaining to the fire command center. Where it is required by other sections or chapters of the 2006 International Building Code, a fire command center for fire-department operations needs to be provided, but before you put it just anywhere in your building or structure, be aware that the location and accessibility of the fire command center must be approved by the fire department. I can tell you that it needs to be separated from the remainder of the building by not less than a 1-hour fire barrier that is constructed in accordance with Chapter 7.

The room for a fire command center has to be a minimum of 96 square feet with a minimum dimension of 8 feet. The layout of the fire command center and all features required by the section have to be submitted for approval; you must submit the layout and get approval before you complete the installation, and the command center has to comply with NFPA 72. The following list contains all the features that are required to be in the fire command center:

- The emergency voice/alarm communication-system unit
- The fire-department communications unit
- Fire-detection and alarm-system annunciator unit
- Annunciator unit visually indicating the location of the elevators and whether they are operational
- Status indicators and controls for air-handler systems
- The firefighter's control panel required by Chapter 9 for smoke-control systems installed in the building
- Controls for unlocking stairway doors simultaneously
- Sprinkler-valve and water-flow detector display panels
- Emergency and standby power status indicators
- A telephone for fire department use with controlled access to the public telephone system
- Fire-pump status indicators
- Schematic building plans indicating the typical floor plan and detailing the building core, means of egress, fire-protection systems, firefighting equipment and fire-department access
- Worktable
- Generator supervision devices, manual start and transfer features
- Public address system, where specifically required by other sections of this code

Fire-department connections must be installed in accordance with the NFPA standard applicable to the system design and must comply with Chapter 9. With respect to hydrants, driveways, buildings, and landscaping, fire-department connections must be located so that fire apparatus and hoses connected to supply the system will not obstruct access to the buildings for other fire apparatus.

The location of the fire-department connections has to be approved before being installed and must be located on the street side of the building, fully visible and recognizable from the street or nearest point of fire-department-vehicle access. On existing buildings, wherever the fire-department connection is not visible to approaching fire apparatus, it must be indicated by an approved sign that has been mounted on the street front or on the side of the building. The sign must have the letters "FDC" at least 6 inches high.

All fire-department connections must be accessible immediately and maintained at all times and without obstruction by fences, bushes, trees, walls, or any other object for a minimum of 3 feet. Keep in mind that the fire-code official is authorized to require locking caps on fire-department connections for water-based fire-protection systems where the responding fire department carries appropriate key wrenches for removal.

A metal sign with raised letters at least 1 inch in size shall be mounted on all fire-department connections serving automatic sprinklers, standpipes, or fire pump connections. Such signs must read "Automatic Sprinklers, Standpipe, or Test Connection" or a combination thereof.

!Codealert

Fire-department connections are covered by Section 912. This section contains numerous code changes that need to be observed.

CHAPTER 10
MEANS OF EGRESS

The provisions of this chapter control the design and construction of means of egress. It is against code to alter a building or structure in a manner that will reduce the number of exits or the capacity of the exits. All means of egress must be maintained in accordance with the International Fire Code. As with past chapters, I have added definition alerts that pertain to parts of the exit code.

GENERAL MEANS OF EGRESS

The general requirements specified in this section apply to all three elements of the exit, way out, or egress (please note that these words may be used interchangeably) system in addition to specific requirements for the exit access, the exit, and the exit discharge detailed elsewhere in this chapter of the code.

At the ceiling construction phase keep in mind there must be a minimum of 80 inches for headroom for any walking surface, including walks, corridors, aisles, and passageways. Not more than 50 percent of the ceiling area of the means of exit can be reduced in height by protruding objects; exceptions to this are door closers and stoppers. The code states that these elements cannot reduce headroom to less than 78 inches, and you have to provide a barrier where the vertical clearance is less than 80 inches high.

The leading edge of any barriers has to be located 27 inches (maximum) from the floor. If you have installed a freestanding object that is mounted on a post or pylon, it cannot overhang more than 4 inches where the lowest point of the leading edge is more than 27 inches and less than 80 inches above the walking surface.

If a sign or other obstruction is mounted between posts or pylons and the clear distance between the posts/pylons is greater than 12 inches, the lowest edge of the sign or obstruction must be 27 inches maximum or 80 inches minimum above the finished floor or ground, except that this requirement does not apply to sloping portions of handrails serving stairs and ramps.

Structural elements, fixtures, or furnishings cannot project horizontally from either side more than 4 inches over any walking surface between the heights of 27 inches and 80 inches above the walking surface. (Handrails that service stair and ramps are permitted to protrude 4.5 inches from the wall.) Protruding objects cannot reduce the minimum clear width of accessible routes, as required in Chapter 11.

Note that all walking surfaces of the means of exit must have a slip-resistant surface and be securely attached. This will protect occupants when traveling from rooms to exits. In those exits where changes in elevations of less than 12 inches exist, sloped surfaces must be used. If the slope is greater than one side or the other, you must install a ramp that complies with this chapter. If the difference in elevation is 6 inches or less, the ramp must be equipped with either handrails or floor-finish materials that contrast with adjacent floor-finish materials. The following is a list of exceptions that pertain to the elevation change:

- A single step with a maximum riser height of 7 inches is permitted for buildings with occupancies in Groups F, H, R-2 and R-3, S, and U at outside doors not required to be accessible by Chapter 11.

- A stair with a single riser or with two risers and a tread is permitted at locations not required to be accessible by Chapter 11, provided that the risers and threads comply with this chapter, have a minimum tread depth of 13 inches, and at least one handrail that complies with this chapter is provided within 30 inches of the centerline of the normal path of exit travel on the stair.

- A step is allowed in aisles serving seating that have a difference in elevation less than 12 inches at locations not required to be accessible by Chapter 11, provided that the risers and treads comply with this chapter and the aisle is provided with a handrail complying with this chapter.

Any change in elevation in a corridor serving nonambulatory persons in Group I-2 occupancy must be by means of a ramp or sloped walkway. The path of the exit cannot be interrupted by any building element other than a means-of-egress compo-

!Codealert

A merchandise pad is an area for display of merchandise surrounded by aisles, permanent fixtures, or walls. Merchandise pads contain elements such as nonfixed and moveable fixtures, cases, racks, counters, and partitions from which customers browse or shop, as indicated in Section 105.2.

> **!Code**alert
> Check Section 1004 for code changes involving occupant load.

nent as specified in this chapter. Obstructions cannot be placed in the required width of a means of exit except projections that are permitted by this chapter. The required capacity of a means-of-exit system cannot be diminished along the path of exit travel. You cannot count elevators, escalators, and moving walkways as a component of a required means of egress from any other part of the building, except when elevators are bused as an accessible means of egress in accordance with this chapter.

OCCUPANT LOAD

When you are constructing a building, you must always use the number of occupants as a basis when determining any means of egress or exit. In areas where occupants from accessory areas exit through a primary space, the calculated occupant load for the primary space must include the total occupant load. Every room or space that is an assembly occupancy must have the occupant load of the room or space posted in a conspicuous place near the main exit or exit-access doorway from the room or space. Posted signs must be of an approved legible permanent design and must be maintained by the owner or authorized agent.

Where exits serve more than one floor, only the occupant load of each floor considered individually must be used in computing the required capacity of the exits at that floor, provided that the exit capacity will not decrease in the direction of exit travel. Where means of egress from floors above and below converge at an intermediate level, the capacity of the means of egress from the point of junction must not be less than the sum of the two floors.

Table 10.1 contains the maximum floor-area allowances per occupant. The number of occupants must be computed at the rate of one occupant per unit of area as in Table 10.1. Where an intended use is not listed in Table 10.1, the building official will establish a use based on a listed use that most nearly resembles the intended use. An exception is that, where approved by the building official, the actual number of occupants for whom each occupied space, floor, or building is designed, although less than those determined by calculation, is permitted to be used in the determination of the design occupant load.

Table 10.1 Maximum floor area allowances per occupancy.

FUNCTION OF SPACE	FLOOR AREA IN SQ. FT. PER OCCUPANT
Accessory storage areas, mechanical equipment room	300 gross
Agricultural building	300 gross
Aircraft hangars	500 gross

Airport terminal
  Baggage claim
  Baggage handling
  Concourse
  Waiting areas
	
	20 gross
	300 gross
	100 gross
	15 gross

Assembly Gaming floors (keno, slots, etc.)	 11 gross
Assembly with fixed seats	See Section 1004.7

Assembly without fixed seats
  Concentrated (chairs only—not fixed)
  Standing space
  Unconcentrated (tables and chairs)
	
	7 net
	5 net
	15 net

Bowling centers, allow 5 persons for each lane including 15 feet of runway, and for additional areas	7 net
Business areas	100 gross
Courtrooms—other than fixed seating areas	40 net
Day care	35 net
Dormitories	50 gross

Educational
  Classroom area
  Shops and other vocational room areas
	
	20 net
	50 net

Exercise rooms	50 gross
H-5 Fabrication and manufacturing areas	200 gross
Industrial areas	100 gross

Institutional areas
  Inpatient treatment areas
  Outpatient areas
  Sleeping areas
	
	240 gross
	100 gross
	120 gross

Kitchens, commercial	200 gross

Library
  Reading rooms
  Stack area
	
	50 net
	100 gross

Locker room	50 gross

Mercantile
  Areas on other floors
  Basement and grade floor areas
  Storage, stock, shipping areas
	
	60 gross
	30 gross
	300 gross

Parking garages	200 gross
Residential	200 gross

Skating rinks, swimming pools
  Rink and pool
  Decks
	
	50 gross
	15 gross

Stages and platforms	15 net
Warehouses	500 gross

For SI: 1 square foot = 0.0929 m^2.

> **!Definition**alert
>
> **Exit:** That portion of a means-of-egress system that is separated from other interior spaces of a building or structure by fire-resistance-rated construction and opening protectives as required, providing a protected path of egress travel between the exit access and the exit discharge. Exits include exterior exit doors at ground level, exit enclosures, exit passageways, exterior exit stairs, exterior exit ramps, and horizontal exits.

For areas that have fixed seats and aisles, the occupant load must be determined by the number of fixed seats installed. The occupant load for areas in which fixed seating is not installed, such as waiting areas and wheelchair spaces, must be determined in accordance with this chapter and added to the number of fixed seats. For areas that have fixed seating without dividing arms, the occupant load cannot be less than the number of seats based on one person for each 18 inches of seating length. The occupant load of seating booths must be based on one person for each 24 inches of booth seat length measured at the backrest of the seating booth.

EGRESS WIDTH

This code requires that the total width of means of egress in inches must not be less than the total occupant load served by the means of egress multiplied by the factors in Table 10.2 and not less than specified elsewhere in this code. Multiple means of egress must be sized such the loss of any one means of egress must not reduce the available capacity to less than 50 percent of the required capacity. The maximum capacity required from any story of a building shall be maintained to the termination of the means of egress.

MEANS-OF-EGRESS ILLUMINATION

The means of egress, including the exit discharge, must be illuminated at all times in the building spaces that the egress occupies, except for the following:

• Occupancies in Group U

• Aisle access in Group A

• Dwelling and sleeping units in Groups R-1, R-2, and R-3

• Sleeping units of Group I occupancies

TABLE 10.2 Egress width per occupant served.

OCCUPANCY	WITHOUT SPRINKLER SYSTEM		WITH SPRINKLER SYSTEM[a]	
	Stairways (Inches per occupant)	Other egress components (inches per occupant)	Stairways (Inches per occupant)	Other egress components (inches per occupant)
Occupancies other than those listed below	0.3	0.2	0.2	0.15
Hazardous H-1, H-2, H-3 and H-4	0.7	0.4	0.3	0.2
Institutional: I-2	N/A	N/A	0.3	0.2

For SI: 1 inch = 25.4. NA = Not applicable.

a. Buildings equipped throughout with an automatic sprinkler system in accordance with Section 903.3.1.1 or 903.3.1.2.

Illumination level cannot be less than 1 footcandle or 11 lux at the walking-surface level except for auditoriums, theaters, concert or opera halls, and similar assembly occupancies; the illumination at the walking-surface level is allowed to be reduced during performances to not less than 0.2 footcandle, provided that the required illumination is automatically restored upon activation of a premises fire-alarm system where such system is provided.

The power supply for exit illumination must normally be provided by the electrical supply. In the event that the power supply is cut off, an emergency electrical system must automatically illuminate the following areas:

• Aisles and unenclosed egress stairways in rooms and spaces that require two or more means of exits

• Corridors, exit enclosures, and exit passageways in buildings required to have two or more exits

!Definitionalert

Accessible Means of Egress: A continuous and unobstructed way of egress travel from any accessible point in a building or facility to a public way.

- Exterior exit components at other than the level of exit discharge until exit discharge is accomplished for buildings required to have two or more exits
- Interior exit-discharge elements, as permitted in this chapter, in buildings required to have two or more exits
- Exterior landings for exit-discharge doorways in buildings required to have two or more exits

The emergency power system must provide power for a duration of not less than 90 minutes and must consist of storage batteries, unit equipment, or an on-site generator. The installation of the emergency power system must be in accordance with Chapter 2.

ACCESSIBLE MEANS OF EGRESS

Accessible spaces must be provided with no less than one means of accessible means of egress. Three exceptions to accessible spaces exist:

- Accessible means of egress are not required in alterations to existing buildings.
- One accessible means of egress is required from an accessible mezzanine level in accordance with this chapter.
- In assembly spaces with sloped floors, one accessible means of egress is required from a space where the common path of travel of the accessible route for access to the wheelchair spaces meets the requirements in this chapter.

Each required accessible means of egress must be continuous to a public way and has to consist of one or more of the following components:

- Accessible routes
- Stairways within vertical exit enclosures
- Exterior exit stairways
- Elevators
- Platform lifts

!Codealert

Egress doors shall be side-hinged swinging, with some exceptions, one of which is a critical or intensive-care patient room within a suite of health-care facilities.

!Definitionalert

Door, Balanced: A door equipped with double-pivoted hardware so designed as to cause a semi-counterbalanced swing action when opening.

- Horizontal exits

- Ramps

- Areas of refuge

These components must comply with Chapter 10 of the 2006 International Building Code. In buildings where a required accessible route is four or more stories above or below a level of exit discharge, at least one required accessible means of egress must be an elevator complying with this section, except for buildings that are equipped throughout with an automatic sprinkler system installed in accordance with Chapter 9. An elevator is not required on floors provided with a horizontal exit located at or above the level of exit discharge or with a ramp that conforms to this chapter.

The use of stairs for an accessible means of exit is common. In order to use stairs in this capacity, the exit stairway has to have a clear width of 48 inches between handrails and must either incorporate an area of refuge within an enlarged floor-level landing or must be accessed from either an area of refuge or a horizontal exit. Please see the following list for exceptions:

- Unenclosed exit stairways as permitted by this chapter are allowed to be considered part of an accessible means of egress.

- The area of refuge is not required at unenclosed exit stairways as permitted by this chapter in buildings or facilities that are equipped throughout with an automatic sprinkler system installed in accordance with Chapter 9.

- The clear width of 48 inches between handrails is not required at exit stairways in buildings or facilities equipped throughout with an automatic sprinkler system installed in accordance with Chapter 9 or for exit stairways accessed from a horizontal exit.

- Areas of refuge are not required at exit stairways serving open parking garages.

In order to be considered part of an accessible means of egress, an elevator must comply with the emergency-operation and signaling-device requirements or Section 2.27 of ASME A17.1, and standby power must be provided in accordance with Chapters 2 and 3. The elevator must be accessed from either an area of refuge complying with this chapter or a horizontal exit; elevators are not required to be accessed from an area of refuge or horizontal exit in open parking garages.

Lifts that are used for wheelchairs do not serve as part of an accessible means of egress, except where allowed as part of a required accessible route in Chapter 11. If you plan on installing a platform lift on an accessible means of egress, it cannot be installed in a fully enclosed or fire-rated shaft.

In your building structure you must have an area of refuge. This refuge space must be accessible from the space it serves by an easily reached means of exit or way out. Keep in mind that this accessible path to the exit has to have a maximum travel distance and cannot be more than the travel distance that is permitted for the type of occupancy that you have built. Every required area of refuge must have a direct access to an enclosed stairway or an elevator.

You may use an elevator lobby for an area of refuge; it must comply with the code regarding smokeproof enclosures. Any refuge, stairway, elevator, and lobby must comply with the code in this chapter. Don't be fooled into thinking that just providing refuge areas is enough. There is a size requirement as well.

Each area of refuge must always be large enough to accommodate one wheelchair space of 30 inches by 48 inches for each 200 occupants, based on the occupant load of the areas of refuge and areas that are served by these areas. The wheelchair spaces cannot reduce the required means of egress width, and access to any of the required wheelchair spaces in an area of refuge can never be obstructed by more than one adjoining wheelchair space.

A separation of each area of refuge by a smoke barrier is mandated. This barrier must comply with Chapter 7, or a horizontal exit complying with this chapter must be provided. Each area of refuge must be designed to minimize the intrusion of smoke, except for those areas located within a vertical exit enclosure. Each area of refuge must have a central control point and must be equipped with a two-way communication system.

If the central control point is not constantly attended, the area has to have controlled access to a public telephone. All locations of the central control point must have a stamp of approval by the fire department. Don't forget that when you choose a two-way communication system, it must have both audible and visible signals. Make sure that any areas of refuge have instructions on the use of the ar-

!Codealert

There are a number of code changes pertaining to exceptions for means of egress.

!Codealert
Stair riser heights shall be 7 inches maximum and 4 inches minimum.
Additional code changes related to stair treads and risers can be found
in Section 1009.3.

eas (under emergency conditions) posted with the communications system. All instructions must have the following items included:

• Directions to other means of egress

• Notification that persons able to use the exit stairway must do so as soon as possible, unless they are assisting others

• Information on planned availability of assistance in the use of stairs or supervised operation of elevators and how to ask for such assistance

• Directions for the use of the emergency communications system

Each door that provides access to an area of refuge from an adjacent floor area
must be identified by a sign that complies with ICC A117.1; it must clearly state
"AREA OF REFUGE" and must include the International Symbol of
Accessibility. In places that require illuminated signs, the area of refuge sign must
be illuminated.

DOORS

The above is just one definition of a door. This section has many rules regarding
doors, gates, and turnstiles, all of which are used as a means of exit. Means-of-exit
doors must be readily distinguishable from the adjacent construction and finishes
such that the doors are easily recognizable as doors.

Mirrors or similar reflecting materials cannot be used on means-of-exit doors.
Means-of-exit doors cannot be hidden by curtains, drapes, decorations, or similar
materials. Have you ever been in a small shop to find that hidden behind a display
is a door? I have, and have often wondered what would happen in an emergency
when the main door will not accommodate the exiting of all the shoppers. This is
exactly what the code is trying to prevent. Let's not forget that the owner of that
shop may be held liable should all shoppers not be able to get outside during such
emergencies.

All exit doors are required to be of a certain size. The minimum width allowed for each door opening has to be sufficient for the occupant load and has to provide a clear width of not less than 32 inches. If the doorway has swinging doors, the clear openings must be measured between the face of the door and the stop, with the door open 90 degrees. Where this section requires a minimum clear width of 32 inches and a door opening includes two door leaves without a mullion (which is a vertical piece of wood that divides the opening), one leaf must provide a clear opening width of 32 inches.

The maximum width of a swinging-door leaf has to be 48 inches nominal. Exit doors in a Group I-2 occupancy used for the movement of beds must have a clear width not less than 41.5 inches. The height of all doors cannot be less than 80 inches. See the following list of exceptions concerning the width and height of doors below:

• The minimum and maximum width does not apply to door openings that are not part of the required means of egress in Group R-2 and R-3 occupancies.

• Door openings to resident sleeping units in Group I-3 occupancies must have a clear width of not less than 28 inches.

• Door openings to storage closets less than 10 square feet in area are not limited by the minimum width.

• Width of door leaves in revolving doors that comply with this chapter are not limited.

• Door openings within a dwelling or sleeping unit cannot be less than 78 inches in height.

• Outside door openings in dwellings and sleeping units, other than the required exit door, are not to be less than 76 inches in height.

• In other than Group R-1 occupancies, the minimum widths apply to inside exit doors within a dwelling or sleeping unit that is not required to be an accessible Type A or Type B unit.

• Door openings required to be accessible within Type B units must have a minimum clear width of 31.75 inches.

!Definitionalert

Horizontal Exit: A path of egress travel from one building to an area in another building on approximately the same level, or a path of egress travel through or around a wall or partition to an area on approximately the same level in the same building, which affords safety from fire and smoke from the area of incidence and areas communication.

For doors that have the required clear width of less than 34 inches above the ground, there must be no projections. Projections into the clear opening width between 34 inches and 80 inches above the floor or ground cannot exceed 4 inches. Did you know that swinging exit doors must be side-hinged? Read on for exceptions to this rule:

• Private garages, office areas, factory areas, and storage areas with an occupant load of 10 or less

• Group I-3 occupancies used as a place of detention

• Critical or intensive-care patient rooms within suites of healthcare facilities

• Doors within or serving a single dwelling unit in Groups R-2 and R-3

• In other than Group H occupancies, revolving doors complying with this chapter

• In other than Group H occupancies, horizontal sliding doors are permitted as a means of egress

• Power-operated doors

• Doors serving a bathroom within an individual sleeping unit in Group R-1

All of these exceptions must comply with their proper chapter or section of the code. Doors must swing in the direction of exit travel where serving an occupant load of 50 or more persons or in a Group H occupancy. The opening force for inside side-swinging doors cannot be more than 5pounds. For other side-swinging, sliding, and folding doors, the door latch must release when subjected to a 15-pound force, and the door must be set in motion when faced with a 30-pound force. The door must swing to a full-open position when subjected to a 15-pound force. All forces must be applied to the side with the latch. If you have installed special doors, such as revolving doors or security grilles, they must comply with the following list:

• Each revolving door has to be capable of collapsing into a book-fold position with parallel exit paths providing an aggregate width of 36 inches.

• A revolving door cannot be located within 10 feet of the foot of or top of stairs or escalators.

• A dispersal area must be provided between stairs or escalators and revolving doors.

• The revolutions per minute for a revolving door cannot be more than those shown in Table 10.3.

• Each revolving door must have a side-hinged swinging door that complies with this chapter in the same wall and within 10 feet of the revolving door.

A revolving door that is used as a component of a means of exit must comply with this chapter and meet the following conditions.

• Revolving doors cannot comprise more than 50 percent of the required exit capacity.

• Each revolving door must be credited with no more than a 50-person capacity.

STAIRWAYS

This section includes stairway topics such as width, headroom treads, and risers and the codes that apply to them. I will also discuss the different types of stairways, such as spiral and aisle stairs. I have also included some information regarding the code for handrails as well. Have you ever used a handrail where your hand just slid down? What if you fell? How can your hand grasp on to the rail for support if the handrail offers no graspability? It is in these instances that the code is put to use, and there are different codes for different aspects of handrails. For instance, handrails with a circular cross-section must have an outside diameter of at least 1.25 inches but no greater than 2 inches or must provide equivalent graspability. If the handrail is not circular, it must have a perimeter dimension of at least 4 inches and not greater than 6.25 inches with a maximum cross-section dimension of 2.25 inches. The maximum radius of edges must be 0.01 inch. Generally speaking, handrail-gripping surfaces must be continuous, without interruption by newel posts or other obstructions.

Stairway widths are an important part of this code. Some types of stairways require different widths, and there are exceptions as well. The code states that the width of stairways cannot be less than 44 inches. There are a few exceptions to this rule:

• Stairways that serve an occupant load of less than 50 must have a width of no less than 36 inches.

• Where an incline platform lift or stairway chairlift is installed on stairways that serve occupancies in Group R-3 or within dwelling units in occupancies in Group R-2, a clear passage width of no less than 20 inches must be provided. If the seat and platform can be folded when not in use, the distance can be measured from the folded position.

All stairways must have a minimum headroom clearance of 80 inches, measured vertically from a line connecting the edge of the nosings. The headroom has to be continuous from above the stairway to the point where the line intersects the landing below, one tread depth beyond the bottom riser. Note, however, that spi-

---

!Definitionalert

Stairway: One or more flights of stairs (a change in elevation, consisting of one or more risers, as defined by the code), either exterior or interior, with the necessary landing and platforms connecting them, to form a continuous and uninterrupted passage from one level to another.

ral stairways, while complying with this section, are allowed to have a 78-inch headroom clearance.

Stair-tread depths have a minimum requirement of 11 inches. Riser heights are measured vertically between the leading edges of adjacent treads, and the tread depth is to be measured horizontally between the vertical planes of the primary projection of adjacent treads and at a right angle to the tread's leading edge. Winders also have a minimum tread depth of 11 inches, measured at a right angle to the tread's leading edge at a point 12 inches from the side where the treads are narrower, and a minimum tread depth of 10 inches. It's important for these measurements to be absolutely accurate. Any difference can make the stairways too steep or appear too high or even too long.

Alternating tread devices, spiral stairways, and aisle stairs are a few exceptions and are discussed later in this chapter. However, in Group R-3 occupancies; within dwelling units in Group R-2 occupancies; and in Group U occupancies that are accessory to a Group R-3 occupancy or accessory to individual dwelling units in Group R-2 occupancies; the maximum riser height is 7.75 inches; the maximum winder-tread depth at the walk line must be 10 inches; and the minimum winder tread depth is 6 inches. A nosing no less than 0.75 inch but not more than 1.25 inches must be provided on stairways with solid risers where the tread depth is less than 11 inches. These measurements are exact and there are no exceptions.

It might seem as if it would be universal knowledge that stair treads and risers must be identical in size and shape, but that is not the case. I've seen stairways in which the treads and risers were of mismatched shape and bigger or smaller than others. Not only does this make the structure look sloppy, but it can be dangerous, depending on the location of the stairway. The tolerance between the largest and smallest riser height or between the largest and smallest tread depth must not be more than 0.375 inch in any flight of stairs.

The greatest winder-tread depth at the 12-inch walk line within any flight of stairs must not exceed the smallest by more than 0.375 inch, measured at a right angle to the tread's leading edge. Where the bottom or top riser adjoins a sloping public way, walkway, or driveway with an established grade and serving as a land-

!Codealert

Curved stairways with winder treads shall have treads and risers in accordance with Section 1009.3, and the smallest radius shall not be less than twice the required width of the stairway, with some exceptions.

!Codealert

Stairways shall have handrails on each side and shall comply with Section 1012. Where glass is used to provide the handrail, the handrail shall also comply with Section 2407 and pertinent exceptions.

ing, the bottom or top riser is permitted to be reduced along the slope to less than 4 inches in height, with the variation in height of the bottom or top riser not to exceed one unit vertical in 12 units horizontal (8-percent slope) of stairway width.

The nosings or leading edges of treads at such nonuniform-height risers must have a distinctive marking stripe, different from any other nosing markings provided on the stair flight. The distinctive marking stripe must be visible when descending the stair and must have a slip-resistant surface. Any marking strip must be a least 1 inch wide but no wider than 2 inches.

Stairway landings must be located at the top and the bottom of each stairway, and the width of such landings cannot be less than the width of the stairway. Every landing must have a minimum dimension measured in the direction of travel equal to the width of the stairway. This dimension does not need to be more than 48 inches where the stairway has a straight run. Please see the following exceptions:

• Aisle stairs complying with the section regarding assemblies

• Doors opening onto a landing cannot reduce the landing to less than one-half the required width. When fully open, the door must not project more than 7 inches into a landing.

Did you know that stairways have to be built of materials that are consistent with the type of construction of the building? Yes, it is true. Wood handrails, however, are permitted for all types of construction. Stairway walking surfaces, of treads, and landings cannot be sloped steeper than 1 unit vertical in 48 units horizontal in any direction. Stairway treads and landing must have a solid surface, and finish floor surfaces must be securely attached. The exception is that in Groups F, H, and S occupancies, other than areas of parking structures accessible to the public, openings in treads and landings are not prohibited, provided that a sphere with a diameter of 1.125 inches cannot pass through the opening.

All outdoor stairways and outdoor approaches to stairways must be designed so that water will not accumulate on walking surfaces. Enclosures under these stairways, such as walls and soffits, must be protected by 1-hour fire-resistance-rated construction or the fire-resistance rating of the stairway enclosure, whichever is

!Codealert

Where the roof-hatch opening providing the required access is located within 10 feet of the roof edge, such roof access or roof edge shall be protected by guards installed in accordance with the provisions of Section 101.3.

greater. Access to the enclosed space cannot be directly from within the stair enclosure; however, spaces under stairways serving and constrained within a single residential dwelling unit in Group R-2 or R-3 are permitted to be protected on the enclosed side with 0.5-inch gypsum board.

Enclosed usable spaces under exterior exit stairways are not permitted unless the space is completely enclosed in 1-hour fire-resistance-rated construction. The open space under exterior stairways cannot be used for any purpose.

A flight of stairs cannot have a vertical rise greater than 12 feet between floor levels or landings, except for aisle stairs that comply with the code. I said earlier that spiral stairways cannot be used as a component in the means of egress. There are a couple of instances in which they are allowed to be used, as follows:

• Spiral stairways are permitted to be used as a component in the means of egress only within dwelling units or from a space no more than 250 square feet in area and serving no more than five occupants or from galleries, catwalks, and gridirons in accordance with the code.

• When a spiral stairway is allowed, there are provisions that must be followed. It must have a 7.5-inch minimum clear tread depth at a point 12 inches from the narrow edge. The risers must be sufficient to provide headroom of 78 inches minimum. Riser height cannot be more than 9.5, inches and the minimum stairway width must be 26 inches.

• Stairways must have handrails on each side and must comply with this chapter of the code.

• Where glass is used to provide the handrail, the handrail must also comply with Chapter 24.

• Aisle stairs complying with this code provided with a center handrail do not need additional handrails.

• Stairways within dwelling units, spiral stairways, and aisle stairs serving seating only on one side are permitted to have a handrail on one side only.

• Decks, patios, and walkways that have a single change in elevation where the landing depth on each side of the change of elevation is greater than what is required for a landing do not require handrails.

• In Group R-3 occupancies, a change in elevation consisting of a single riser at an entrance or egress door does not require handrails.

• Changes in room elevations of only one riser within dwelling and sleeping units in Group R-2 and R-3 occupancies do not require handrails.

If your building or structure has four or more stories in height above grade plane, one stairway must extend to the roof surface, unless the roof has a slope steeper than 4 units vertical in 12 units horizontal (33 percent slope). In buildings without an occupied roof, access to the roof from the top story must be permitted by an alternating tread device.

RAMPS

The provisions of this section apply to ramps used as a component of a means of egress. Please see the list of exceptions below:

• Other than ramps that are part of the accessible routes providing access or ramped aisles within assembly rooms

• Curb ramps must comply with ICC A117.1

• Vehicle ramps in parking garages not used for pedestrian exit access do not have to comply with this section when they are not part of an accessible route serving accessible parking spaces, or as a means of exit

Ramps that are used as part of a means of egress must have a running slope no steeper than 1 unit straight up and 12 units across. The slope of the other pedestrian ramps cannot be steeper than one unit straight up in eight units horizontal, except for an aisle ramp in occupancies of Group A. Ramps must have landings at the bottom of each ramp and at points of turning, entrance, exits, and doors. The slope of a ramp's landing cannot be steeper than one unit vertically, or a 2 percent slope in any direction. You are not permitted to make any changes in these levels.

!**Definition**alert

Ramp: A walking surface that has a running slope steeper than 1 unit vertical in 20 units horizontal with a 5 percent slope.

EXIT SIGNS

Doors, stairways, and ramps without exit signs may just appear to lead to another level or another room in the building or structure. All exits and exit-access doors have to be marked by an approved exit sign that can be seen from any direction. The sign must be placed in an area so that no point in a corridor is more than 100 feet from the nearest visible exit area. Exceptions are as follows:

• Exit signs are not required in rooms or areas that require only one exit or exit access.

• Main exterior exit doors or gates that are obviously and clearly identifiable as exits need not have exit signs where approved by a building official.

• Exit signs are not required in occupancies in Group U and individual sleeping units or dwelling units in Group R-1, R-2, or R-3.

• Exit signs are not required in sleeping areas in occupancies in Group I-3.

• In occupancies in Groups A-4 and A-5, exit signs are not required on the seating side of a stadium or theater or for openings into seating areas so long as exit signs are provided in the concourse that are readily apparent from the seating area in a stadium or theater in an emergency.

Every sign, including directional signs, has to have legible letters without decoration. Letters must be at least 6 inches high and no less than 0.75 inches wide. Spacing between letters may not be less than 0.375 inches. Larger signs must have letter widths and spacing in proportion to their height.

HANDRAILS

The code for handrail extensions has changed from previous code. According to the 2006 International Building Code, stair risers have a maximum height requirement of 7 inches and a minimum of 4 inches. What good are stairways and ramps without handrails? A handrail is defined as a horizontal or sloping rail intended for

!Codealert

The floor or ground surface of a ramp run or landing shall extend 12 inches minimum beyond the inside face of a handrail complying with Section 1012.

grasping by the hand for guidance or support. Handrails must be sufficiently attached and adequate in strength in accordance with Chapter 16 and be no less than 34 inches but not more than 38 inches high.

Handrail height is to be measured from above the stair-tread nosing. When used on ramps, handrail height is measured from the surface of the ramp slope. Some handrails have a circular cross-section, which must have an outside diameter of at least 1.25 inches but not greater than 2 inches, or must be able to be grasped easily. If the handrail is not circular, it must have a perimeter dimension of at least 4 inches but not greater than 6.25 inches with a maximum cross-section dimension of 2.25 inches. Any edge has to have a minimum radius of 0.01 inch. Gripping surfaces of all handrails is to be continuous, without interruption by newel posts or any other obstruction. Please see the following list for exceptions.

- Handrails within dwelling units are permitted to be interrupted by a newel post at a stair landing.
- A volute, turnout, or starting ease is allowed on the lowest tread in houses only.
- Handrail brackets or balusters that are attached to the bottom surface of the handrail and do not hang over the sides of the handrail within 1.5 inches are not considered obstructions.

There must be a clear space between a handrail and a wall or other surface that must be at least 1.5 inches. Keep handrails, walls, or other surfaces adjacent or the handrail free of any sharp or abrasive elements. Projections into the required width of stairways and ramps cannot be more than 4.5 inches at or below the handrail height.

GUARDS

A guard is a building component or a system of building components located at or near the open sides of elevated walking surfaces that minimizes the possibility of a fall from the walking surface to a lower level. Guards must be placed along the sides of stairways, ramps, and landings that are more than 30 inches above the floor or grade below where the glazing does not meet the strength requirements in Chapter 16. There are places where guards are not required:

- On the loading side of docks or piers
- On the side of stages (from the audience side) and raised platforms, including steps leading up to the stages and raised platforms
- At vertical openings in the performance area of stages and platforms
- At elevated walking surfaces attached to stages and platforms providing access to and utilization of special lighting or equipment
- Along vehicle service pits not accessible to the public
- In assembly seating where guards in accordance with this chapter are permitted and provided

Guards are used to perform a protective barrier, but the purpose is defeated if the barriers are not of a certain height. According to the code, a guard cannot be less than 42 inches high and must be measured vertically above the leading edge of the tread, next to the walking surface or beside a seat. However, for occupancies in Group R-3 and within individual housing units in occupancies in Group R-2, guards whose top rail also serves as a handrail must have a height of no less than 34 and no more than 38 inches, also measured vertically from the leading edge of the stair tread and nose.

The height in assembly seating areas must be in accordance with this chapter. Guards are in play to prevent you, clients, customers, and the people close to you from getting hurt. Therefore, they must be installed in screened porches and decks where the walking surface is more than 30 inches above the floor or grade below.

Any component of mechanical equipment that requires service, such as appliances, equipment, fans, or roof-hatch openings located within 10 feet of a roof edge or open side of a walking surface and more than 30 inches above the floor or roof requires the installation of a guard that prevents the passage of a 21-inch object.

EXIT ACCESS

An exit access is the part of the means-of-egress system that leads from any occupied portion of a building or structure to an exit. This can be an aisle, hallway, or other means of a walkway. Egress or throughway cannot pass through rooms that are of high-hazard occupancy unless the adjoining rooms are accessory to the area and in Groups H, S, or F occupancy when the adjoining or intervening rooms or spaces are the same or a less hazardous occupancy group.

An exit cannot pass through kitchens, storage rooms, closets, or other spaces that are used for similar purposes. Keep in mind that an exit is not allowed to pass through a room that can be locked or through bedrooms or bathrooms. When more than one tenant lives on any one floor of a building or structure, each tenant space, dwelling, or sleeping unit must be provided with access to the required exits in a clear aisle that does not pass through another tenant unit.

In Group I-2 occupancies, such as hospitals or nursing facilities, rooms or suites must have an exit-access door that leads directly to a corridor. Please see the following list for exceptions:

• Rooms with exits doors that lead or open directly to the outside at the ground level are exempt.

• Patient sleeping rooms are permitted to have one intervening room if that room is not used as an exit access for more than eight patients.

• Special nursing suites are allowed to have one intervening room if they allow direct and constant (visual) supervision by nursing personnel.

!Codealert

The word "EXIT" shall be in high contrast with the background and shall be clearly discernible whether the means of exit-sign illumination is or is not energized. If a chevron directional indicator is provided as part of the exit sign, the construction shall be such that the direction of the chevron directional indicator cannot be readily changed.

• Any room that is not located in a suite where the travel distance is greater than 100 feet is allowed to have exit travel.

In occupancies other than Groups H-1, H-2, and H-3, the common path of egress or exit travel is not allowed to be more than 75 feet; Groups H-1, H-2, and H-3 cannot have a common path of egress travel of more than 25 feet. In Group B, F, and S occupancies exit travel cannot be more than 100 feet, and that maximum is allowed only if the building is equipped with an automatic sprinkler system. If your tenant space, which is considered a Group B, S, or U occupancy, has an occupant load of not more than 30 people, the length of a common path travel cannot be more than 100 feet.

If your building or structure is of Group I-3 occupancy, the path of travel is limited to 100 feet. And finally, the length of a common path in Group R-2 occupancy cannot be more than 125 feet, provided you have installed an automatic sprinkler system throughout. Always remember that any automatic sprinkler systems have to be installed in accordance with this code.

Aisles

Aisles that serve as part of the exit access in a means-of-egress system must comply with this section. Aisles must be provided from all occupied parts of the exit access that contain seats, tables, furnishings, displays, and similar fixtures or equipment. The aisles must be unobstructed for the full required width. Doors, when fully opened, and handrails must not reduce width by more than 7 inches. And doors in any position must not reduce the required width by more than one-half. Other projections such as trim and similar decorative features are allowed to project into the required width by 1.5 inches on each side.

For Group M occupancies where seating is located at a table or counter and is adjacent to an aisle or aisle accessway, the measurement must be made to a line of

!Definitionalert

Aisle: An exit-access component that defines and provides a path of egress travel.

at least 19 inches. This must be measured from the edge of the table or counter. In the case of other side boundaries for aisle or aisle accessways, the clear width must be measured to walls, edges of seating, and tread edges, except that handrail projections are permitted.

If tables or counters have fixed seating, meaning that the seats are bolted to the floor, the width of the aisle accessway must be measured from the back of the chair. Table and seating accessways are required to have a minimum of 12 inches of width plus 0.5 inches of width for each additional foot. The exception to this is a part of an aisle accessway not more than 6 feet long and used by no more than four people. Balconies used for egress purposes have to conform to the same requirements as corridors.

EXIT-ACCESS DOORWAYS

Requirements for the number of exits and exit-access doorways are dependant upon a number of details in regard to occupancy. Take a look at Table 10.4 for more information.

All required exits must be located in such a way that their use is obvious. Interior exit stairways and ramps must be enclosed with fire barriers or horizontal assemblies that comply with Chapter 7 of the code. A fire-resistance rating of no less than 2 hours is required when the unit is four stories or more and no less than 1 hour when the unit is less than four stories. Be sure to include basements in this calculation but not mezzanines. Although obvious, an exit enclosure cannot be used for any purpose other than a means of egress. The exceptions are included in the list below:

• In all occupancies other than Group H and I, a stairway is not required to be enclosed when the stairway serves an occupant load of less than 10 and the stairway is open to not more than one story above or below the story at the level of exit discharge. In all cases, however, the maximum number of connecting open stories cannot be more than two.

• In buildings of Group A-5 where all portions of the means of egress are essentially open to the outside need not be enclosed.

!Codealert

Section 1014, dealing with exit access, has many recent code updates that must be considered.

- Stairways that are used and contained in a single residential dwelling or sleeping unit in Group R-1, R-2, or R-3 occupancies are not required to be enclosed.
- Stairways that are not a required means-of-egress element are not required to be enclosed where they comply with Chapter 7.
- Stairways in open parking structures that serve only the parking structure are not required to be enclosed.
- Stairways in Group I-3 occupancies and means-of egress-stairways as provided for in Chapter 4 are not required to be enclosed.
- In groups other than Group H and Group I occupancies, a maximum of 50 percent of egress stairways serving one adjacent floor are not required to be enclosed, as long as at least two means of egress are provided from both floors that are using the open stairways. Any floors that are interconnected cannot be open to other floors.
- In groups other than Group H and I occupancies, inside stairway exits that provide service for the first and second stories of a building that has an automatic sprinkler system throughout are not required to be enclosed, but only if there are at least two means of exits. These interconnected floors cannot be open to other stories.

TABLE 10.4 Spaces with one means of egress.

OCCUPANCY	MAXIMUM OCCUPANT LOAD
A, B, E[a], F, M, U	49
H-1, H-2, H-3	3
H-4, H-5, I-1, I-3, I-4, R	10
S	29

a. Day care maximum occupant load is 10.

Any penetrations into and any openings through an exit enclosure are not allowed except for required exit doors, equipment, and ductwork that is necessary for independent pressurization, sprinkler piping, standpipes, and electrical raceways serving the exit enclosure and terminating at a steel box that is not more than 16 square inches in size. Such penetrations must be protected in accordance with Chapter 7. There must not be any penetrations or communication openings, whether protected or not, between adjacent exit enclosures. Equipment and ductwork for exit-enclosure ventilation as allowed by this chapter must meet the following conditions:

- All equipment and ductwork has to be located outside the building and must be directly connected to the exit enclosure by ductwork enclosed in construction as required by shafts.
- Where the equipment and ductwork is located within the exit enclosure, the intake air must be taken directly from the outdoors and the exhaust air must be discharged directly to the outdoors, or the air must be conveyed through ducts enclosed in construction as required by shafts.
- Where equipment and ductwork is located within the building, this equipment and ductwork must be separated from the rest of the building, including other mechanical equipment, with construction as required for shafts.

In each case, openings into the fire-resistance-rated construction must be limited to those that are needed for maintenance and operation and must be protected by opening protectives in accordance with Chapter 7 for shaft enclosures. And all exit-enclosure ventilation systems must be independent of other building ventilation systems.

The outside walls of an exit enclosure must comply with the requirements of Chapter 7 for exterior walls. In cases where nonrated walls or unprotected openings enclose the outside of the stairway and the walls or openings are exposed by other parts of the building at an angle of less than 180 degrees, the outside building walls within 10 feet horizontally of a nonrated wall or unprotected opening must have a fire-resistance rating of no less than 1 hour. And as a reminder, openings within inside walls must be protected by opening protectives that have a fire-rotection rating of no less than 3/4 hour. This type of construction is to extend vertically from the ground to a point that is 10 feet above the uppermost landing of the stairway or to the roof line, whichever isis the lower level.

!Definitionalert

Emergency Escape: An operable window, door, or other device that provides for means of escape and access for rescue in the event of an emergency.

!Codealert
Where access to three or more exits is required, at least two exit doors or exit-access doorways shall be arranged in accordance with the provisions of Section 1015.2.1.

FLOOR NUMBER SIGNS

Earlier in this chapter, I talked about sign requirements. All floor landings in every inside exit enclosure that connects more than three stories must have a floor number sign. This sign must identify the floor level and the direction to the exit of that stairway. It must also include the availability of the roof access from the stairway for the fire department and be located 5 feet above the floor landing is such a position that it is visible when the doors are in the open and closed positions.

SMOKEPROOF ENCLOSURES

A smokeproof enclosure or pressurized stairway must exit into a public space or into an exit passageway, yard, or open space. Any of these must have direct access to a public area. An exit passageway has to have no other openings and must be separated from the remainder of the building by 2-hour fire-resistance-rated construction. Please see the following list of exceptions:

• Openings in the exit passageway that serves a smokeproof enclosure are allowed where the exit passageway is protected and pressurized in the same manner as the smokeproof enclosure and openings are protected as required for access from other floors.

• Openings in the exit passageway serving a pressurized stairway are permitted where the exit passageway is protected and pressurized in the same manner as the pressurized stairway.

• A smokeproof enclosure or pressurized stairway is allowed to pass through areas on the level of discharge or vestibules as allowed by this chapter.

The access to the stairway must be through an entrance hall or an open balcony, unless the pressurized stairway complies with Chapter 9.

HORIZONTAL EXITS

A horizontal exit cannot serve as the only exit from a portion of a building. In buildings where two or more exits are required, not more than one-half of the total number or width of exits can be horizontal exits. However, if your building is of Group I-2 occupancy, horizontal exits are permitted to make up two-thirds of the required exits.

In Group I-3 occupancies, horizontal exits are allowed to make up 100 percent of the exits required. There must at least 6 square feet of accessible space per occupant on each side of the exit for the total number of people in adjoining compartments. Horizontal exits that connect separations between buildings or refuge areas must provide a firewall or a fire barrier complying with Chapter 7 and having a fire-resistance rating of no less than 2 hours. Such exit separation must extend vertically through all levels of the building unless the floor assemblies have a fire-resistance rating of no less than 2 hours, with all openings having a protective element. The exception is that a fire-resistance rating is not required at horizontal exits between a building area and an above-grade pedestrian walkway. The walkway has to provide a distance of more than 20 feet and must comply with Chapter 31 on special construction.

Horizontal exits that contain a refuge area must be in a space occupied by the same tenant or public area, and each refuge area has to be big enough to accommodate the original occupant load plus any extra people from the adjoining compartment. The anticipated occupant load from the adjoining compartment is based on the capacity of the horizontal-exit doors that enter the refuge area. The net floor area allowable per occupancy must be as follows for the indicated occupancies:

• Six square feet per occupant for occupancies in Group I-3.

• Fifteen square feet per occupant for ambulatory occupancies and thirty square feet per occupant in nonambulatory occupancies in Group I-2.

EXTERIOR EXIT RAMPS AND STAIRWAYS

If your building or structure includes outside exit ramps and stairways that are used as part of a required means of exit, pay close attention to this section. If your building or structure is of Group I-2 occupancy, you may not use exterior ramps and stairways as an element of a means of exit.

In any other occupancy buildings cannot exceed six stories above grade plane or have occupied floors more than 75 feet above the lowest level of fire-department access. All exterior exit ramps and stairways must have at least one open side and a minimum of 35 square feet of total open area bordering each floor level and the level of each intermediate landing. Take note that the required open area must be located no less than 42 inches above the bordering floor or landing level.

!Codealert

See Section 1020, on vertical exit enclosures, for code changes and additions.

These types of ramps and stairways have to be separated from the inside of the building. Remember, any opening is limited to those necessary for exit from spaces that are normally occupied. However, there are several exceptions to this rule:

• Separation from the interior of the building is not required for occupancies, other than those in Group R-1 or R-2, in buildings that are no more than two stories above grade plane where the level of exit discharge is the first story above grade plane.

• Separation from the interior of the building is not required where the exterior ramp or stairway is served by an exterior ramp and/or balcony that connects two remote outside stairways or other approved exits with a perimeter that is no less than 50 percent open. For the opening to be considered open, it must be a minimum of 50 percent of the height of the enclosing wall, with the top of the openings less than 7 feet above the top of the balcony.

• Separation from the inside of the building is not required for an inside ramp or stairway located in a building or structure that is allowed to have unenclosed inside stairways in accordance with the chapter.

• Separation from the inside of the building is not required for exterior ramps or stairways connected to open-ended corridors, provided that these conditions are met: the building, which includes corridors and ramps and/or stairs, must be thoroughly equipped with an automatic sprinkler system; and the open-ended corridors must be connected on each end to an outside exit ramp or stairway and comply with this chapter.

• At any location in an open-ended corridor where a change of direction exceeds 45 degrees, a clear opening of no less than 35 square feet or an outside ramp or stairway must be provided. Ifn clear openings are provided; they must be located so that the accumulation of smoke or toxic gases is minimized.

ASSEMBLY

Assemblies, in regard to occupancies in Group A, contain seats, tables, displays, equipment, or other materials and must comply with this section. Please note that bleachers, grandstands, and folding and telescopic seating must comply with ICC 300. Any Group A occupancy that has an occupant load of 300 or more must provide a main exit and must be wide enough to accommodate at least one-half of the occupant load. The width cannot be less than the total width that is required of all means of egress that leads to the exit.

If a building or structure is classified as Group A occupancy, the main exit has to have frontage on at least one street or an unoccupied space that cannot be less than 10 feet in width and that borders on a street or public way.

Buildings or structures that are classified as assembly occupancies and that do not have any well-defined main exits or that have multiple main exits are allowed to be circulated around the perimeter of the building, provided that the total width of the egress is not less than 100 percent of the required width. Also, you must remember that in addition to having access to a main exit, each level in Group A occupancies that has an occupancy load over 300 must have additional means of exits that have the capacity for at least one-half of the total occupant level. Again, where there is no well-defined exit, you may distribute the exits around the perimeter as long as the total width of the egress is not less than 100 percent of the required width.

Let's say you have a Group A-1 occupancy and people enter your building, but there is no available seating and you allow them to wait in the lobby or foyer. You must keep the lobby or foyer space separated from the means of egress by partitions or by a fixed rigid railing. The railing cannot be less than 42 inches high. A foyer, if not directly connected to a public street by the main entrance or exit, must have a straight and clear corridor or path to the main entrance or exit. If your Group A occupancy has an interior balcony or gallery and the seating capacity is 50 or more, it, too, must provide two means of egress, one from each side of every balcony or gallery, and at least one leading directly to an exit.

The clear width of the means of egress for smoke-protected assembly seating cannot be less than the occupant load served by the egress element multiplied by the appropriate factor in Table 10.5. In Group A assemblies the travel distance between exits and aisles must be located in such a way that distance is not more than 200 feet measured along the line of travel in nonsprinklered buildings. If the building is equipped with sprinklers, the travel distance is 250 feet. If aisles are provided for seating, the distance is measured along the aisle accessway without travel over or on the seats. See the following list for exceptions:

• For smoke-protected assembly seating the travel distance from each seat to the nearest entrance to an opening cannot be more than 200 feet. The travel distance from the entrance to the opening to a stair, ramp, or walkway on the outside of the building cannot be more than 200 feet.

• For open-air seating the travel distance from each seat to the buildings outside cannot be more than 400 feet. The travel distance is not limited in facilities of Type I or II construction.

Any occupant of a seat cannot walk more than 30 feet before coming to a point where a choice of two exit pathways exists. If the occupant load is less than 50, the common path of travel cannot be more than 75 feet, and for smoke-protected assembly seating, the common path of travel cannot be more than 50 feet. Note that this is a change from the previous code. Each of these aisles must end at a cross aisle, foyer, doorway, opening, or area with access to an exit.

Dead-end aisles must not be greater than 20 feet in length; however, they are allowed where seats beyond 20 feet are no more than 24 seats from another aisle. This measurement must be made along a row of seats with a minimum clear width of 12 inches plus 0.06 inch for each additional seat above seven in a row. For smoke-protected assembly seating, the dead-end aisle length of vertical aisles cannot be more than a distance of 21 rows. This aisle cannot be more than 40 seats from another aisle, measured along a row of seats having an aisle accessway with a minimum clear width of 12 inches plus 0.3 inch for every additional seat above seven in a row. There can be no obstructions in the required width of aisles except for handrails. Table 10.5 shows exceptions for smoke-protected assembly seating.

EMERGENCY ESCAPE AND RESCUE

This is the final section of Chapter 10. It discusses the provisions that must be made for emergency escape and rescue in Group R and I-1 occupancies.

Basements and sleeping rooms that are located below the fourth story above grade plane must have at least one outside emergency escape and rescue opening. If a basement contains one or more bedrooms, emergency egress and openings are required for each bedroom but are not required in adjoining areas of a basement. A list of exceptions follows:

• Occupancies other than Group R-3 equipped throughout with an approved automatic sprinkler system do not require emergency egress.

• Occupancies other than Group R-3 that have rooms used for sleeping and have a door that leads to a fire-resistance-rated corridor that has access to two remote exits in opposite directions do not need emergency egress.

!Codealert

With some exceptions, the common path of egress travel shall not exceed 30 feet from any seat to a point where an occupant has a choice of two paths of egress travel to two exits.

• The emergency escape and rescue opening is permitted to open onto a balcony, provided the balcony provides access to an exit and the dwelling or sleeping unit has an exit that does not open to an atrium.

• Basements that have a ceiling height of less than 80 inches or basements without habitable space and no more than 200 square feet in floor area are not required to have an emergency escape and rescue window.

• High-rise buildings in accordance with Chapter 4 do not need emergency egress.

All emergency escape and rescue openings must have a minimum net clear opening of 5.7 square feet, except for grade-floor openings, which must be at least 5 square feet; both require minimum height of 24 inches and width of 20 inches. The maximum height from the floor in all cases is to be no greater than 44 inches measured from the floor. Keep in mind that all the emergency openings do not provide safety if they do not work right or cannot be opened from the inside. You can place bars, grilles, or grates over the escape, but they must be removable from the inside without the use of a tool or key. If you choose to install any of the above on an emergency escape, you must also install smoke alarms.

To conclude this chapter, it is important that you realize that these provisions are in play for a number of reasons, with safety being one of the most important. As with any chapter of this book, any questions can be addressed to your local code office or building inspector.

TABLE 10.5 Smoke-protected assembly aisle accessways.

TOTAL NUMBER OF SEATS IN THE SMOKE-PROTECTED ASSEMBLY OCCUPANCY	MAXIMUM NUMBER OF SEATS PER ROW PERMITTED TO HAVE A MINIMUM 12-INCH CLEAR WIDTH AISLE ACCESSWAY	
	Aisle or doorway at both ends of row	Aisle or doorway at one end of row only
Less than 4,000	14	7
4,000	15	7
7,000	16	8
10,000	17	8
13,000	18	9
16,000	19	9
19,000	20	10
22,000 and greater	21	11

For SI: 1 inch = 25.4 mm.

CHAPTER 11
ACCESSIBILITY

This chapter is dedicated to accessible design and construction for physically disabled persons. All buildings and facilities are to be designed and constructed to be accessible according to the provisions of the IBC and ICC A117.1. I have included definition alerts and tables for added information. Do not assume that the code is the same as in previous codes; there have been changes.

SCOPE OF ACCESSIBILITY REQUIREMENTS

The following lists places and sites where accessibility is required:

• Employee work areas must be designed and constructed so that individuals with disabilities can approach, enter, and exit the work area.

• Work areas or portions that are less than 150 square feet in area and are elevated 7 inches or more above the ground or finish floor are exempt from all requirements, provided that the elevation is essential to the function of that space.

• Walk-in coolers and freezers intended for employee use only are not required to be accessible.

• Occupancies in Group U are exempt from the requirements of this chapter other than the following: in agricultural buildings, access is required to paved work areas and areas open to the general public. Private garages or carports that require accessible parking.

• In buildings where a-day care facility (Groups A-3, E, I-4, and R-3) is part of a dwelling unit, only the portion of the structure used for the facility is required to be accessible.

• The operable parts on fuel-dispensing devices must comply with ICC A177.1, Section 308.2.1, or 308.3.1.

The next list contains sites and places where accessibility rules are not required. This list and the one above are summaries only and are not intended to take

!**Definition**alert

Accessible: A site, building, facility, or portion thereof that complies with this chapter of the IBC.

the place of accessibility requirements for routes, entrances, and other provisions that will be discussed later in this chapter:

• Detached one- and two-family dwelling and accessory structures and their associated sites and facilities are not required to be accessible.

• Construction sites, structures, and equipment directly associated with the actual processes of construction including but not limited to scaffolding, bridging, materials, hoists, or construction trailers are not required to be accessible.

• Raised areas used primarily for purposes of security, life safety, or fire safety including but not limited to observation galleries, prison-guard towers, fire towers, or lifeguard stands are not required to be accessible or to be served by an accessible route.

• Limited-access spaces such as nonoccupiable spaces accessed only by ladders, catwalks, crawl spaces, freight elevators, or very narrow passageways are not required to be accessible. This includes but is not limited to spaces frequented only by personnel for maintenance, repair, or monitoring of equipment such as elevator pits or water- or sewage-treatment pump rooms.

• Single-occupant structures accessed only by passageways below grade or elevated above grade including but not limited to toll booths that are reached only by underground tunnels are not required to be accessible.

• Buildings of Group R-1 containing not more than five sleeping units for rent or hire that are also occupied as the residence of the proprietor are not required to be accessible.

!**Code**alert

A dwelling unit or sleeping unit must comply with this code and the provisions for accessible units in ICC A 117.1.

!Codealert

Accessible units and Type B units shall be provided in Group R-4 occupancies in accordance with Sections 1107.6.4.1 and 1107.6.4.2.

• In detention and correctional facilities, common-use areas that are used only by inmates or detainees and security personnel and that do not serve holding cells or housing cells required to be accessible are not required to be accessible or to be served by an accessible route.

ACCESSIBLE ROUTES

Accessible routes within a site must be provided from public-transportation stops; accessible parking; accessible passenger-loading zones, and public streets or sidewalks to buildings' accessible entrance. Other than buildings or facilities or serving Type B units, an accessible route is not required between site arrival points and the building or facility entrance if the only means of access between them is a vehicular way not providing for pedestrian access.

When a building or portion of a building is required to be accessible, an accessible route must be provided to each portion of the building and to accessible building entrances that connect each accessible pedestrian walkway and the public way. However, in assembly areas with fixed seating required to be accessible, an accessible route is not required to serve fixed seating where wheelchair spaces or designated aisle seats required to be on an accessible route are not provided.

Connected areas and walkways include employee work areas, which must have accessible common-use circulation paths. See the following list of exceptions for common-use circulation paths:

!Definitionalert

Accessible Route: A continuous, unobstructed path that complies with this chapter of the IBC.

• Any path located within work areas that are less than 300 square feet and have definite installed partitions, counters, casework, or furniture is not required to be an accessible route.

• Any paths that are an integral component of equipment transportation are not required to be accessible routes.

• Work areas that are fully exposed to the weather are not required to have accessible routes.

Let's say that your building or structure is a multilevel facility. Did you know that you need to have at least one accessible route connecting each accessible level? It's true, and the same rule applies to mezzanines. There are, however, many exceptions to multilevel buildings, so please read the following paragraph carefully and see which ones apply to you.

An accessible route is not required to stories and mezzanines above and below accessible levels that have a total area of not more than 3,000 square feet. This exception does not apply to the following:

• Multiple tenant facilities of Group M occupancies containing five or more tenant spaces.

• Levels containing offices of healthcare providers.

• Passenger transportation facilities and airports in Group A-3 or B occupancies.

In Groups A, I, R, and S occupancies, levels that do not contain accessible elements or other spaces as required by this chapter are not required to be served by an accessible route from an accessible level. If your building is an air-traffic-control tower or a two-story building that has an occupant load of five or less people (as long as it does not contain public areas), you are not required to have an accessible route. As you can see, there are many buildings and structures that do not require accessible routes. Please be sure to ask your local code or building official if you have any questions regarding these codes. Let's move on and look at the different locations in which an accessible route is required by the 2006 International Building Code.

Accessible routes must occur together with or be located in the same area as a general-circulation path. Where the circulation path is inside, the accessible path

---

!Codealert

All or any portion of a space used only by employees and only for work. Corridors, toilet rooms, kitchenettes, and break rooms are not employee work areas.

!Codealert

With one exception, at least one wheelchair space shall be provided in team or player seating areas serving areas of sport activity.

must be inside as well. If you have only one accessible route in a building, it cannot pass through kitchens, storage rooms, restrooms, closets, or similar spaces. Please note that the only time a single accessible route may pass through a kitchen or storage room is in an accessible Type A or Type B dwelling unit.

Accessible routes from parking garages in Type B dwelling units are not required to be located inside. If you have installed security barriers that include but are not limited to security posts and security check points, they are not allowed to obstruct a required accessible route or accessible means of egress. If the security barriers incorporate elements that cannot comply with these requirements, such as certain metal detectors or other similar devices, the accessible route is allowed to provide adjacent security screening devices.

Any persons with disabilities who have to pass through one of these devices must be able to maintain visual contact with their belongings to the same extent as others do. One example of where this rule might apply is at an airport-security checkpoint when baggage is being checked while going through an X-ray machine. People with disabilities must be able to have the same ability as others to watch their belongings pass through.

ACCESSIBLE ENTRANCES

In addition to the provisions in this chapter regarding accessible entrances, at least 60 percent of all public entrances must be accessible. This does not apply to areas

!Definitionalert

Public Entrance: An entrance that is not a service entrance or a restricted entrance.

that are not required to be accessible and loading and service entrances that are not the only entrance to a tenant space. Please see the following list for public entrances that must be accessible:

• Parking garage entrances: direct access for pedestrians from parking structures to building or facility entrances must be accessible.

• Entrances from tunnels or elevated walkways: if direct access is provided for pedestrians from a tunnel or elevated walkway to a building or facility, at least one entrance from each tunnel or walkway must be accessible.

• Restricted entrances: in situations where restricted entrances are provided to a building or facility, at least one such entrance to the building must be accessible.

• Entrances for inmates or detainees: of entrances used only by inmates or detainees and security personnel at judicial facilities, detention facilities, or correctional facilities, at least one must be accessible.

• Service entrances: if a service entrance is the only entrance to a building or a tenant space in a facility, that entrance must be accessible.

• Tenant spaces, dwelling units, and sleeping units: at least one accessible entrance must be provided to each tenant, dwelling unit, and sleeping unit in a facility, except that it is not required for Type A or Type B dwelling units and sleeping units that are not required to be accessible units.

PARKING AND PASSENGER LOADING FACILITIES

For buildings and structures that provide parking, accessible parking spaces must be provided in compliance with Table 11.1, except as required by other sections of this chapter. In cases where more than one parking facility is provided on a site, the number of parking spaces required to be accessible must be calculated separately for each parking facility. Note that this section does not apply to parking spaces that are used exclusively for buses, trucks, other delivery vehicles, law enforcement, or motor pools where lots accessed by the public are provided with an accessible passenger-loading zone. The next paragraphs will give detailed information regarding the accessible percentages that different groups or facilities must provide for parking.

!Codealert

A dwelling unit or sleeping unit must comply with this code and the provisions for accessible units in ICC A 117.1.

TABLE 11.1 Accessible parking spaces.

TOTAL PARKING SPACES PROVIDED	REQUIRED MINIMUM NUMBER OF ACCESSIBLE SPACES
1 to 25	1
26 to 50	2
51 to 75	3
76 to 100	4
101 to 150	5
151 to 200	6
201 to 300	7
301 to 400	8
401 to 500	9
501 to 1,000	2% of total
1,001 and over	20, plus one for each 100, or fraction thereof, over 1,000

For Groups R-2 and R-3 Type A or B dwelling or sleeping units, 2 percent, but not less than 1, of each type of parking space provided for occupancies are required to be accessible. In locations where parking is provided within or beneath a building, accessible parking spaces must also be provided. Hospital outpatient facilities must provide 10 percent of accessible patient and visitor parking spaces for hospital outpatient facilities. For rehabilitation facilities and outpatient physical-therapy facilities, 20 percent, and not less than one, of the portion of patient and visitor parking spaces must be accessible.

!Codealert
Accessible units shall be provided in Group I-3 occupancies in accordance with Sections 1107.5.5.1 through 1107.5.5.3.

!Codealert

Accessible units and Type B units shall be provided in general-purpose hospitals, psychiatric facilities, detoxification facilities, and residential-care/assisted-living facilities of Group I-2 occupancies in accordance with Sections 1107.5.3.1 and 1107.5.3.2.

A percentage of accessible parking spaces must be dedicated to van-accessible parking. At least one must be a van-accessible parking space. There are provisions regarding the location of these spaces. Accessible parking spaces must be located on the shortest accessible travel route from adjacent parking to an accessible building entrance. For parking facilities that are not provided for any particular building, accessible parking spaces must also be located on the shortest route to an accessible pedestrian entrance. For buildings that have multiple accessible entrances with adjacent parking, accessible parking spaces must be dispersed and located near the accessible entrances. Pay attention to the following exceptions:

• In multilevel parking structures, van-accessible parking spaces are permitted on one level.

• Accessible parking spaces are allowed to be located in different parking facilities if substantially equal or greater accessibility is provided in terms of distance from an accessible entrance or entrances, parking fee and user convenience.

DWELLING AND SLEEPING UNITS

In addition to the other requirements of this chapter, occupancies with dwelling or sleeping units must be provided with accessible features in accordance with this section. Dwelling and sleeping units that are required to be accessible units and

!Codealert

A parcel of land bounded by a lot line or a designated portion of a public right-of-way is a site.

!Codealert

A Type A unit is a dwelling unit or sleeping unit designed and constructed for accessibility in accordance with this code and the provisions for Type A units in ICC A117.1.

Type A and B units must comply with the applicable portions of Chapter 10 of
ICC A117.1. Rooms and spaces that are used by the general public or available for
use by residents must be accessible. Some examples of these rooms or spaces are:

• Bathrooms

• Kitchens

• Living and dining rooms

• Patios

• Terraces and balconies

This list of required rooms does not include recreational facilities in accordance with this chapter. The primary entrance of each of these accessible units, not
rooms, must have at least one accessible route connecting the building or facility
entrances. However, sometimes there are circumstances beyond your control.
Assume that either the slope of the finished ground level between the accessible
facility and the building is more than one vertical unit in 12 or if physical barriers
or legal restrictions prevent the installation of an accessible route. You are allowed, in accordance with the 2006 International Building Code, to install a vehicular route (with parking that complies with this chapter) at each public or commonly used facility or building in place of the accessible route.

!Codealert

A dwelling unit or sleeping unit with habitable space located on more
than one story is a multistory unit.

> # !Codealert
>
> A Type B unit is a dwelling unit or sleeping unit designed and constructed for accessibility in accordance with this code and the provision for Type B units in ICC A117.1, consistent with the design and construction requirements of the federal Fair Housing Act.

There have been a number of changes for Group I occupancies. At least 4 percent, but not less than one, of the dwelling and sleeping units must be accessible units. Group I-1 structures of Type B units that contain four or more dwelling or sleeping units intended to be occupied as a residence must be accessible, and the number of Type B units is allowed to be reduced in accordance with this chapter. Group I-2 occupancies, nursing homes, must have at least 50 percent (but not less than one) accessible and Type B units. Accessible and Type B units must be provided for Group I-2, which include the following:

• General-purpose hospitals

• Psychiatric facilities

• Detoxification facilities

• Residential-care/assisted-living facilities

Of all the dwelling and sleeping units in these facilities, at least 10 percent, and no less than one, must be accessible units. For structures with four or more dwelling or sleeping units with the intention to be occupied as a residence, every unit must be a Type B occupancy. However, the number of Type B units is permitted to be reduced in accordance with this chapter. Group I-2 is classified as hospitals and rehabilitation facilities. These facilities must specialize in treating conditions that affect mobility or contain units that do specialize in such treatment, and 100 percent of these dwelling units must be accessible. The above code provisions pertain to Group I occupancies only.

> # !Codealert
>
> At least 50 percent, but not less than one, of each type of the dwelling and sleeping units shall be accessible units.

Apartment houses, monasteries, and convents belong to Group R-2. Both Type A and B must be provided for this group. If this particular group contains more than 20 percent dwelling or sleeping units, at least 2 percent must be Type A, and all units on site will need to be considered when determining the total number of units. Two exceptions to this are: (1) the number of Type A units is permitted to be reduced and (2) existing structures on a site do not have to be included when calculating the total number of units.

In cases where there are four or more dwelling or sleeping units intended to be occupied as a residence, they must be Type B units. Again, the number of Type B units is allowed to be reduced in accordance with this chapter. Not all Group R-2 occupancies are apartment houses, monasteries, and convents. Any other occupancy that fall under Group R-2 must provide accessible and Type B units.

The required number of Type A and Type B units do not apply to a site where the required elevation of the lowest floor or the lowest horizontal structural building members of non-elevator buildings are at or above the design flood elevation, which results in the following:

• A difference in elevation between the minimum required floor elevation at the primary entrances and vehicular and pedestrian arrival points within 50 feet over 30 inches

• A slope exceeding 10 percent between the minimum required floor elevation at the primary entrances and vehicular and pedestrian arrival points within 50 feet

SPECIAL OCCUPANCIES

In addition to other requirements of this chapter, this code section must be applied to special occupancies. Theaters, bleachers, grandstands, stadiums, arenas, and other fixed-seating-assembly areas must provide accessible wheelchair spaces that comply with ICC A117.1. If such stadiums and the like provide audible public announcements, they must also provide equivalent text information regarding events and facilities. For example, places or areas that have electronic signs must have the ability to display text that is being announced over a PA system, with the

!Codealert

Where required, sites, buildings, structures, facilities, elements, and spaces, temporary or permanent, shall be accessible to persons with physical disabilities.

!Codealert

Walk-in coolers and freezers intended for employee use only are not required to be accessible.

exception of announcements that cannot be prerecorded. Please note that Table 11.2 contains information regarding capacity seating and the minimum number of wheelchair spaces that must be provided.

Team or player seating at a sports activity must provide at least one wheelchair space, except for team or player seating serving bowling lanes. Another special occupancy that has provisions for accessibility is a performance area. An accessible route must directly connect the performance area to the assembly seating area. Self-service storage areas must provide accessible individual self-storage space in accordance with Table 11.3.

The final provisions for special occupancies that I want to go over are judicial facilities. Each courtroom and central holding cell must be accessible. Where separate central holding cells are provided for adult male and juvenile males and adult and juvenile females, one of each type must be accessible. In places where central holding cells are provided but are not separated by age or gender, at least one accessible cell has to be provided. This same is true in courtrooms.

TABLE 11.2   Accessible wheelchair spaces.

CAPACITY OF SEATING IN ASSEMBLY AREAS	MINIMUM REQUIRED NUMBER OF WHEELCHAIR SPACES
4 to 25	1
26 to 50	2
51 to 100	4
101 to 300	5
301 to 500	6
501 to 5,000	6, plus 1 for each 150, or fraction thereof, between 501 through 5,000
5,001 and over	36 plus 1 for each 200, or fraction thereof, over 5,000

TABLE 11.3 Accessible self-service storage facilities.

TOTAL SPACES IN FACILITY	MINIMUM NUMBER OF REQUIRED ACCESSIBLE SPACES
1 to 200	5%, but not less than 1
1,001 and over	10, plus 2% of total number of units over 200

OTHER FEATURES AND FACILITIES

A very important area that must adhere to the 2006 International Building Code accessibility provisions is bathrooms and bathing areas. Not only does the code provide for accessibility, but for separate facilities for both genders. All bathrooms and bathing facilities must be made accessible.

Where a floor level is not required to be connected by an accessible route, the only bathroom or bathing room provided cannot be located on the inaccessible floor. You must provide at least one of each type of fixture, element, control, or dispenser in each accessible bathroom/bathing room. There are several exceptions to these provisions:

• Bathrooms or bathing facilities accessed only through a private office and not intended for common or public use are allowed to have doors that swing into the clear floor space, provided the door swing can be reversed.

• The height requirements for the water closet in ICC A117.1 are not applicable.

• Grab bars are not required, and the requirement for height, knee, and toe clearance does not apply to a lavatory.

• Where multiple single-user toilet rooms or bathing facilities are clustered at a single location, at least 50 percent but not less than one room for each use at each cluster must be accessible.

!Codealert
Group I-2 nursing homes require accessible units, and Type B units shall be provided in accordance with Sections 1107.5.2.1 and 1107.5.2.2.

!Codealert
Where drinking fountains are provided on an exterior site, on a floor or within a secured area, the drinking fountains shall be provided in accordance with Sections 1109.5.1 and 1109.5.2.

The provisions for unisex toilet rooms state that they must include only one water closet and only one lavatory. A unisex bathing room in accordance with this chapter will be considered a unisex toilet room. A urinal is allowed in addition to the water closet in a unisex toilet room. Unisex bathing rooms can include only one shower or bathtub fixture and only one water closet and lavatory. An accessible route is mandatory and must be not be more than 500 feet from any separate-sex bathroom.

PLATFORM LIFTS

Platform lifts are allowed to be part of a required accessible route in new construction where indicated in the following list:

- An accessible route to a performing area and speaker platforms in Group A occupancies
- An accessible route to wheelchair spaces required to comply with the wheelchair-space-dispersion requirements of this chapter
- An accessible route to spaces not open to the general public with an occupant load of not more than five
- An accessible route within a dwelling or sleeping unit
- An accessible route to wheelchair seating spaces located in outdoor dining terraces in Group A-5 occupancies where the means of egress from the dining terraces to a public way are open to the outdoors
- An accessible route to the following: jury boxes, witness stands, raised courtroom stations, judges' benches, and stations for clerks, bailiffs, deputies, and court reporters
- An accessible route to load and unload areas serving amusement rides
- An accessible route to play components or safe contained play structures
- An accessible route to team or player seating areas serving areas of sport activity

> ## !Codealert
> Lawn-seating areas and exterior-overflow seating areas, where fixed seats are not provided, shall connect to an accessible route.

• An accessible route where existing exterior-site constraints make use of a ramp or elevator infeasible

Take note that some of the items in the above list are changes from the previous code and deserve special attention.

SIGNAGE

This section focuses on the accessible elements that are also required by the International Symbol of Accessibility. There are several locations where signs are mandated. Such locations are noted below:

• Accessible parking spaces required by this chapter except where the total number of parking spaces provided is four or less
• Accessible passenger loading zones
• Accessible areas of refuge
• Accessible rooms where multiple singe-user toilet or bathing rooms are clustered at a single location
• Accessible entrances where not all entrances are accessible
• Accessible check-out aisles where not all aisles are accessible--the sign, where provided, must be above the check-out aisle
• Unisex bath and bathing rooms

> ## !Codealert
> See Section 1109.13 and Section 1109.14 for recent changes in the code.

• Accessible dressing, fitting, and locker rooms where not all such rooms are accessible

Other signs indicating special accessibility provisions are located at each door to an egress stairway, exit passageway, and at all areas of refuge.

CHAPTER 14
EXTERIOR WALLS

In this chapter you will find discussion of the following topics:

- Exterior walls
- Exterior wall coverings
- Exterior wall openings
- Exterior windows and doors
- Architectural trim
- Balconies and similar projections
- Bay and oriel windows

PERFORMANCE REQUIREMENTS

This section applies to exterior walls, wall coverings, and their components. All exterior walls need to be weather-protected with a weather-resistant exterior-wall envelope.

An exterior-wall envelope must include flashing and must be designed and constructed in such a manner that an accumulation of water within the wall assembly is prevented. The envelope must provide a water-resistive barrier behind the exterior veneer and a way for any water that enters the assembly to the exterior to drain. All protection against condensation in the exterior-wall assembly must be provided in accordance with the International Energy Conservation Code.

There are exceptions to weather protection: a weather-resistant exterior-wall envelope is not required over concrete or masonry walls that are designed in accordance with Chapters 19 and 21, and compliance with the requirements for a means of drainage is not required for an exterior-wall envelope that has been tested and proven to resist wind-driven rain. This includes joints, penetrations, and

> **!Definition**alert
>
> **Exterior-Wall Envelope:** A system or assembly of exterior-wall compo-
> nents, including exterior-wall finish materials, that provides protection of
> the building structural members, including framing and sheathing materials,
> and conditioned interior space from the detrimental effects of the exterior
> environment.

intersections with different materials in accordance with ASTM E 331 but only
under the following conditions:

• Exterior-wall-envelope test assemblies must include at least one opening, one
  control joint, one wall/eave interface, and one wall sill.

• All tested openings and penetrations must be representative of the intended end-
  use configuration.

• Exterior-wall-envelope test assemblies must be at least 4 feet by 8 feet (1,219
  mm by 2,438 mm) in size.

• Exterior-wall-envelope assemblies must be tested at a minimum differential
  pressure of 6.2 pounds per square feet.

• Exterior-wall-envelope assemblies must be subjected to a minimum test exposure duration of 2 hours.

Exterior-wall design has to resist wind-driven rain, and the results of such testing have to indicate that no water could penetrate any control joints in the wall,
joints at the perimeter of openings, or intersections or terminations with different
materials. The structural aspects of exterior walls and associated openings must be
designed in accordance to Chapter 16 on structural design. Please refer to Chapter
7 for fire-resistance-rating requirements for exterior walls.

If your building or structure is in a flood zone, it is required that exterior walls
extend below the design-floor elevation and be resistant to water damage. Wood
must be treated by a pressure preservative in accordance with AWPA U1 for the
species, product, and end use using a preservative that is found in Section 4 of
AWPA U1 or decay-resistant heartwood, redwood, black locust, or cedar. Please
make note that the provisions for wood represent a change from the previous code.
In flood-hazard areas that are subject to high-velocity wave action, electrical, mechanical, and plumbing components cannot be mounted on or penetrate through
exterior walls that are designed to break away under flood loads.

MATERIALS

Any materials that you use in the construction of exterior walls have to comply with this section of the code. You may use materials not listed in this section, but be sure that any other materials or alternatives have been approved by the building official. If not approved, you will have wasted your time, energy, and money. A simple phone call will give you the answers to your questions regarding other materials that you wish to use. When constructing a water-resistive barrier, a minimum of one layer of asphalt felt, which must be No. 15, complying with ASTM D 226 for Type 1 felt or other approved materials, must be attached to the studs or sheathing. This has to include flashing to provide a continuous water-resistive barrier behind the exterior-wall veneer.

If you have chosen to construct exterior walls with cold-rolled copper, the copper needs to conform to the requirements of ASTM B 370 and all lead-coated copper must conform to the requirements of ASTM B 101. These requirements are a change from the previous code. Please refer to the following list for other materials and the code chapter with which they must be in accordance:

- Concrete and glass-unit masonry: exterior walls of concrete construction must be designed and constructed in accordance with Chapter 21 on masonry.
- Plastics: plastic panel, apron, or spandrel walls as defined by the code are not limited in thickness, provided that plastic and their assemblies conform to the requirements of Chapter 26 on plastic, and are constructed of approved water-resistant materials of adequate strength to resist the wind loads for cladding specified in Chapter 16 on structural design.
- Vinyl siding must be certified and labeled as conforming to the requirements of ASTM D 3679 by an approved quality- control agency.
- Fiber-cement siding must conform to the requirements of ASTM C 1186 and be identified on the label listing as being tested by an approved quality-control agency.

As you can see, there are specific requirements that materials must adhere to. Not only does brick or vinyl siding add to the beauty of a building, but it serves as the first line of defense against the elements of the weather.

!Definitionalert

Veneer: A facing attached to a wall for the purpose of providing ornamentation, protection, or insulation but not adding strength to the wall.

FLASHING

Table 14.1 covers the minimum thickness of weather coverings. Exterior walls must provide weather protection for buildings, and in Table 14.1 you will find the acceptable coverings that are approved for weather protection.

Flashing must be installed in such a manner so as to prevent moisture form entering the wall or to redirect it to the outside. The following list contains all of the perimeters in which flashing must be installed:

• Perimeters of exterior-door and -window assemblies

• Penetrations and terminations of exterior-wall assemblies

• Exterior-wall intersections with roofs, chimneys, porches, decks, balconies, and similar projections

• Built into gutters and similar locations

Flashing with projection must be installed on both sides and the ends of copings, under sills, and continuously above projecting trim. Moisture is also known to build up in exterior-wall pockets and even crevices of buildings and structures. It is crucial that these wall pockets be avoided or protected by using caps or drips. Any means of preventing water damage can be used so long as you obtain approval by the building official.

Are you familiar with where flashing and weepholes must be located? If you said in the first course of masonry above the finished ground level, give yourself a star. This is correct, and it includes the level above the foundation wall or slab. There are other points of support, including structural floors, shelf angles, and lintels, where anchored veneers must be designed in accordance with this section.

VENEERS

There are several types of veneers that you need to be familiar with; the following section describes these types and the provisions that they must follow. Wood veneers that are found on exterior walls of buildings of Type I, II, III, and IV con-

!Definitionalert

Flashing: Pieces of sheet metal or the like, used to reinforce and weatherproof the joints and angles of a roof where it comes in contact with a wall.

TABLE 14.1 Minimum thickness of weather coverings.

COVERING TYPE	MINIMUM THICKNESS (inches)
Adhered masonry veneer	0.25
Aluminum siding	0.019
Anchored masonry veneer	2.625
Asbestos-cement boards	0.125
Asbestos shingles	0.156
Cold-rolled copper[d]	0.0216 nominal
Copper shingles[d]	0.0162 nominal
Exterior plywood (with sheating)	0.313
Exterior plywood (without sheating)	See Section 2304.6
Fiber cement lap siding	0.25[c]
Fiber cement panel siding	0.25[c]
Fiberboard siding	0.5
Glass-fiber reinforced concrete panels	0.375
Hardboard siding[c]	0.25
High-yield copper[d]	0.0162 nominal
Lead-coated copper[d]	0.0216 nominal
Lead-coated high-yield copper	0.0162 nominal
Marble slabs	1
Particleboard (with sheating)	See Section 2304.6
Particleboard (without sheating)	See Section 2304.6
Precast stone facing	0.625
Steel (approved corrosion resistant)	0.0149
Stone (cast artificical)	1.5
Stone (natural)	2
Structural glass	0.344
Stucco or exterior portland cement plaster	
Three-coat work over:	
Metal plaster base	0.875[b]
Unit masonry	0.625[b]
Cast-in-place or precast concrete	0.625[b]
Two-coat work over:	
Unit masonry	0.5[b]
Cast-in-place or precsat concrete	0.375[b]
Terra cotta (anchored)	1
Terra cotta (adhered)	0.25
Vinyl siding	0.035
Wood shingles	0.375
Wood siding (without sheating)[a]	0.5

For SI: 1 inch = 25.4 mm.

a. Wood siding of thicknesses less than 0.5 inch shall be placed over sheating that conforms to Section 2304.6.

b. Exclusive of textures.

c. As measured at the bottom of decorative grooves.

d. 16 ounces per square foot for cold-rolled copper and lead-coated copper, 12 ounces per square foot for copper shingles, high-yield copper and lead-coated high-yield copper.

!Codealert

Vinyl siding is a shaped material, made principally from rigid polyvinyl chloride (PVC), that is used as an exterior wall covering.

struction cannot be less than 1 inch thick. Exterior hardboard siding cannot be less than 0.438 inch and must conform to the following:

* The veneer cannot be more than three stories in height, measured from the grade plane.

* Where fire-retardant-treated wood is used, the height cannot be more than four stories.

* The veneer is attached to or furred from a noncombustible backing that is fire-resistance-rated as required by other provisions of the code.

* Where open or spaced wood veneers (without concealed spaces) are used, they cannot project more than 24 inches from the building wall.

When using anchored masonry veneer, it is important that you remember that not only does it have to comply with the IBC but also with ACI 530/ASCE 5 and TMS 402. Anchored masonry veneers in accordance with this chapter are not required to meet the tolerances in Article 3.3 G1 or ACI 530.1/ASCE 6/TMS 602, but slab-type veneer units no more than 2 inches in thickness must be anchored directly to masonry, concrete, or stud construction.

If the veneer units are of marble, granite, or other stone units of slab and are hung from ties of corrosion-resistant dowels in drilled holes, they must be located in the middle third of the edge of the units. The spacing of these units cannot be more than 24 inches maximum. Each unit cannot have less than four ties per veneer unit and cannot be more than 20 square feet in area.

If you choose to use veneer ties made of metal, make sure they are smaller in area than 0.0336 by 1 inch or, if made of wire, not smaller in diameter than 0.1483 inch. Some people prefer anchored terracotta or ceramic units on exterior walls of their buildings. These, too, have to be installed according to code restrictions.

Whether terracotta is tied or not, the minimum thickness is 1.625 inches and it must be directly anchored to masonry, concrete, or stud construction. Tied terracotta or ceramic veneer units, along with the minimum thickness of 1.625 inches, must also have projecting dovetail webs on the back surface spaced approximately every 8 inches. Veneer ties must have sufficient strength to support the full weight

!**Code**alert

A material behind an exterior wall covering that is intended to resist liquid water that has penetrated behind the exterior covering from further intrusion into the exterior-wall assembly is a water-resistive barrier.

of the veneer in tension. Adhered masonry veneer must comply with the applicable requirements in Section 1405.9.1 and Sections 6.1 and 6.3 of ACI 530/ASCE 5/TMS 402.

Metal supports for exterior metal veneer must be protected. You can paint, galvanize, or use another coating or approved treatment. Wood studs, furring strips, or other wood supports for exterior metal veneer must be approved and pressure-treated. Joints and edges that are exposed to the weather have to be caulked with a durable waterproofing material to prevent moisture from seeping in. Be sure that you choose an approved material. Grounding of metal veneers on buildings must comply with the requirements of Chapter 27 of the code or the ICC Electrical Code.

When using the next type of veneer, glass, there are a number of guidelines you must follow. The area of a single section of thin exterior structural-glass veneer cannot be more than 10 square feet in cases where it is not more than 15 feet above the level of the sidewalk or grade level below and cannot be more than 6 square feet where it is more than 15 feet above that level.

Neither the length nor the height of the glass can be more than 48 inches, nor can the glass be any thicker than 0.344 inch. Only apply the glass veneer after making sure that the backing is thoroughly dry. You must apply an approved bond coat as well. Make sure this is applied uniformly over the entire surface of the backing so as to effectively seal the surface. The glass can then be set in place with mastic cement so that at least 50 percent of the area of each glass unit is directly bonded to the backing by mastic not less than 0.25 inches thick and not more than 0.625 inches thick. The mastic cement must be of an approved kind, and make note that the bond coat and mastic must be evaluated for compatibility and must adhere firmly together.

In areas where glass extends to a sidewalk surface, each section must rest in an approved metal molding and be set at least 0.25 inches above the highest point of the sidewalk. The space between the molding and the sidewalk must be thoroughly caulked and made water-tight. If set above sidewalk level and/or above 36 inches, the mastic-cement binding must be supplemented with approved nonferrous-metal shelf angles located in the horizontal joints in every course. These shelf angles cannot be less than 0.0478 inch thick and no less than 2 inches long and must be spaced at approved intervals, with no less than two angles for each glass unit.

Securing angles to the wall or backing should be done with expansion bolts, toggle bolts, or another approved method. If you decide you want to use glass veneers, realize that there are specifications for the use of joints. Unless otherwise specifically approved by the building official, abutting areas must be ground square and mitered joints cannot be used except where specifically approved for wide angles.

All glass veneers must be held in place by the use of fastenings at each vertical or horizontal edge or at the four corners of each glass unit, securing the fastenings to the wall or backing with expansion bolts, toggle bolts, or other methods. The exposed edges of structural glass veneer must be flashed with overlapping corrosion-resistant metal flashing and caulked with a waterproof compound to prevent moisture from coming between the glass veneer and the backing.

WINDOW SILLS

Where the opening of the sill portion of an operable window is located more than 72 inches above the finished grade, the lowest part of the clear opening of the window must be a minimum of 24 inches above the finished floor surface of the room in which the window is located. This applies to the following list of occupancies:

• Group R-2

• Group R-3

• One-family

• Two-family

• Multiple-family

Glazing between the floor and a height of 24 inches must be fixed or have openings that will not allow a 4-inch-diameter sphere to pass through. One exception to this is an opening provided with no window guard that complies with ASTM F 2006 or F 2090.

!Definitionalert

Vinyl Siding: A shaped material, made principally from rigid polyvinyl chloride (PVC), that is used as an exterior wall covering.

ject to cyclical flexural response due to wind loads cannot demonstrate any major loss of tensile strength. At one time or another you may have seen buildings that have gravel and stone on top of their roofs. But did you realize that buildings located in hurricane alleys or regions are forbidden to have gravel or stone on their roofs? The height of a building is also taken into consideration for gravel and stone. Table 15.1 provides the regulations regarding the maximum allowance of roof height permitted for buildings with gravel or stone on the roof in areas outside a hurricane-prone region.

FIRE CLASSIFICATION

Roof assemblies are divided into classes A, B, and C. The minimum roof coverings installed on buildings must comply with Table 15.2 based on the type of construction of the building. An exception is that skylights and sloped glazing must comply with Chapter 24.

TABLE 15.1 Maximum allowable mean roof height permitted for buildings with gravel or stone on the roof in areas outside a hurricane-prone region.

| BASIC WIND SPEED FROM FIGURE 1609 (mph)[b] | MAXIMUM MEAN ROOF HEIGHT (ft)[a,c] | | |
| | Exposure category | | |
	B	C	D
85	179	60	30
90	110	35	15
95	75	20	NP
100	55	15	NP
105	40	NP	NP
110	30	NP	NP
115	20	NP	NP
120	15	NP	NP
Greater than 120	NP	NP	NP

For SI: 1 foot = 304.8 mm, 1 mile per hour = 0.447 m/s.
a. Mean roof height in accordance with Section 1609.2.
b. For intermediate values of basic wind speed, the height associated with the next higher value of wind speed shall be used, or direct interpolation is permitted.
c. NP = gravel and stone not permitted for any roof height.

!Definitionalert

Roof Assembly: A system designed to provide weather protection and resistance to design loads. The system consists of a roof covering and roof deck or a single component serving as both. A roof assembly includes the roof deck, vapor retardant, substrate or thermal barrier, insulation, and roof covering.

Let's take a look at the different classes and their requirements. Class A roof assemblies are those that are most effective against severe fire exposure. All Class A roof assemblies and coverings must be listed and identified as such by an approved testing agency. You may use this class for all buildings or structures of all types of construction, except for Class A roof assemblies with the following coverings:

- Brick
- Masonry
- Slate
- Clay
- Concrete roof tile
- Exposed-concrete roof deck
- Ferrous or copper shingles or sheets

TABLE 15.2 Minimum roof covering classification for types of construction.

IA	IB	IIA	IIB	IIIA	IIIB	IV	VA	VB
B	B	B	C^c	B	C^c	B	B	C^c

For SI: 1 foot = 304.8 mm, 1 square foot = 0.0929 m^2.

a. Unless otherwise required in accordance with the International Wildland-Urban Interface Code or due to the location of the building within a fire district in accordance with Appendix D.

b. Nonclassified roof coverings shall be permitted on buildings of Group R-3 and Group U occupancies, where there is a minimum fire-separation distance of 6 feet measured from the leading edge of the roof.

c. Buildings that are not more than two stories in height and having not more than 6,000 square feet of projected roof area and where there is a minimum 10-foot fire-separation distance from the leading edge of the roof to a lot line on all sides of the building, except for street fronts or public ways, shall be permitted to have roofs of No. 1 cedar or redwood shakes and No. 1 shingles.

Class B roof assemblies are those that are effective against moderate fire-test exposure; they must be listed and identified as such by an approved testing agency. An exception is Class B roof assemblies with coverings of metal sheets and shingles. Class C roof assemblies are those that are effective against light fire-test exposure. Class C assemblies must also be listed and identified by an approved testing agency.

MATERIALS

The requirements in this section apply to the application of roof-covering materials. When choosing your roofing materials, it is a requirement that all materials must be compatible with one another and with the building or structure and the materials to which they are applied.

There are standards that materials must conform to (applicable standards). If the materials are of questionable suitability, testing by an approved agency is required by the building official to determine their quality or any limitations of application.

If you are receiving your roofing materials in bulk shipments, be aware that all materials must have all information issued by the manufacturer in the form of a certificate or on a bill. They must also be in packages that have the manufacturer's identifying marks and approved testing-agency labels.

ROOF COVERINGS

Requirements for roof coverings include their method of application. Pay attention to the manufacturer's installation instructions. For roofs located in areas where the basic wind speed is 110 mph or greater, asphalt shingles must be tested in accordance with ASTM D 3161, Class F.

Asphalt shingles must have self-seal strips or be interlocking. All packaging must have a label indicating compliance or a listing by an approved testing

!Codealert

See Section 1504.8 for changes in code requirements for gravel and stone.

!Definitionalert

Underlayment: One or more layers of felt, sheathing paper, nonbituminous saturated felt, or other approved material over which a steep-slope roof covering is applied.

agency. Don't think that you can use any type of fasteners for this type of shingle. All fasteners for asphalt shingles must be galvanized stainless-steel, aluminum, or copper roofing nails. These nails must have a minimum 12-gauge shank with a minimum diameter head of 0.375 inch.

If roof sheathing is less than 0.75 inch, you must be sure that the nail will penetrate. The code requires that there be a minimum number of fasteners per shingle--no less than four fasteners per strip shingle or two fasteners per individual shingle.

There has been a change in the code regarding the underlayment application for roof slopes.

For roof slopes of 4 units vertical in 12 units horizontal or greater, underlayment must be in one layer applied shingle fashion, parallel to and starting from the eave, lapped 2 inches, and fastened sufficiently to hold in place.

Distortions in the underlayment must not interfere with the ability of the shingles to seal. In areas of high winds (over 110 mph), underlayment must be applied with corrosion-resistant fasteners along the overlap at a maximum spacing of 36 inches on center, making sure to follow the manufacturer's instructions.

As you are aware, various exterior-building elements such as roofs, walls, windows, and doors form a packet that protects the inside of the building from weather elements, water being the biggest threat. Obviously a roof with appropriate drainage minimizes this threat, but buildings need more than just a roof. This is why the installation of flashing is so important. Installed at intersecting roofs and parapets, this corrosion-resistant metal overlaps to discourage water entrapment.

Base flashing is the portion of the installation that is attached to the roof itself, and cap flashing is attached to the projection or wall. Both must be made of a corrosion-resistant metal.

Valleys occur where the different roof slopes intersect and are considered to be problem spots. Valley linings are to be installed in accordance with the manufacturer's instructions before applying shingles. For open valleys lined with metal, the lining must be at least 16 inches wide. Lining of two plies of mineral-surfaced roll roofing is permitted, with the bottom layer having a maximum of 18 inches and the top layer a minimum of 36 inches.

For closed valleys (valleys covered with shingles), lining of just one ply of smooth roll roofing at least 36 inches wide is permitted. Table 15.3 provides the requirements for valley lining materials.

Be sure to provide a drip edge at eaves and gables of shingle roofs. Overlap to a minimum of 2 inches and extend 0.25 inches below sheathing. Drip edges must be mechanically fastened a maximum of 12 inches.

When you have a tall projection or structure mounted on a pitched roof, such as a chimney, you must use a saddle or cricket. A cricket is actually a ridge that is installed between the roof slope and the structure. Cricket or saddle coverings must be made of sheet metal or of the same material as the roof covering.

Another type of roofing, mineral-surfaced roll roofing, cannot be applied on roof slopes below 1 unit vertical in 12 units horizontal. You must be careful of the underlayment of this type of roofing (refer to ASTM D 226, Type II, Class M mineral-surfaced roll roofing).

TABLE 15.3 Valley lining material.

MATERIAL	MINIMUM THICKNESS	GAGE	WEIGHT
Aluminum	0.024 in.	—	—
Cold-rolled copper	0.0216 in.	—	ASTM B 370, 16 oz. per sq. ft.
Copper	—	—	ASTM B 370, 16 oz. per sq. ft.
Galvanized steel	0.0179 in.	26 (zinc-coated G90)	—
High-yield copper	0.0162 in.	—	ASTM B 370, 16 oz. per sq. ft.
Lead	—	—	2.5 pounds
Lead-coated copper	0.0216 in.	—	ASTM B 101, 16 oz. per sq. ft.
Lead-coated high-yield copper	0.0162 in.	—	ASTM B 101, 12 oz. per sq. ft.
Painted terne	—	—	20 pounds
Stainless steel	—	28	—
Zinc alloy	0.027 in.	—	—

For SI: 1 inch = 25.4 mm, 1 pound = 0.454 kg, 1 ounce = 28.35 g.

!Definitionalert

Penthouse: An enclosed, unoccupied structure above the roof of a building, other than a tank, tower, spire, dome cupola, or bulkhead, occupying not more than one-third of the roof area.

In areas where there is a history of ice formation, an ice barrier that consists of at least two layers of underlayment cemented together or of a self-adhering polymer-modified bitumen sheet must be used instead of normal underlayment; it must extend from the eave's edge to a point at least 24 inches inside the exterior wall line of the building. This does not include detached accessory structures that contain no conditioned floor area. The last type of roof covering is wood shingles.

INSULATION

Although short, this section includes a table that provides the required material standards for roof insulation. The use of above-deck thermal insulation is permitted, provided the insulation is covered with an approved roof covering that passes the tests of FM 4450 or UL 1256. This does not include foam-plastic roof insulation, which must conform to the material and installation requirements found in Chapter 26, or to above-deck thermal insulation covered with an approved roof covering for a concrete deck. If you are choosing to install cellulosic-fiberboard roof installation, you must be sure that it conforms to the material and installation requirements of Chapter 26. All above-deck thermal insulation board must comply with the standards in Table 15.4.

STRUCTURES

A penthouse or other projection above the roof in structures of other than Type I construction cannot be more than 28 feet above the roof, where used as an enclosure for tanks or for elevators that run to the roof, and not more than 18 feet for all other cases. A penthouse, bulkhead, or any other similar projection above the roof cannot be used for any other purpose other than housing for mechanical equipment or shelter for vertical shaft openings in the roof.

If a penthouse or bulkhead is being used for purposes other than those permitted by this section, the use must conform to the requirements of the code. This sec-

TABLE 15.4 Material standards for roof insulation.

Cellular glass board	ASTM C 552
Composite boards	ASTM C 1289, Type III, IV, V or VI
Expanded polystyrene	ASTM C 578
Extruded polystyrene board	ASTM C 578
Perlite board	ASTM C 728
Polyisocyanurate board	ASTM C 1289, Type I or Type II
Wood fiberboard	ASTM C 208

tion does not prohibit the placing of wood flagpoles on the roof of any building. The type of construction for penthouses must include walls, floors, and roofs as required for that building type. Exceptions to this include the following:

- On buildings of Types I and II construction, the exterior walls and roofs of penthouses with a fire-separation distance of more than 5 feet and less than 20 feet must be of at least 1-hour fire-resistance-rated noncombustible construction.
- Interior framing and walls must be made of noncombustible construction.
- On buildings of Types III, IV, and V construction, the exterior walls of penthouses with a fire-separation distance of more than 5 feet and less than 20 must be at least 1-hour fire-resistance-rated construction.
- Walls with a fire-separation distance of 20 feet or greater from a common property line must be of Type IV or noncombustible construction.
- Interior framing and walls must be Type IV or noncombustible construction.
- Unprotected noncombustible enclosures housing only mechanical equipment and located with a minimum fire-separation distance of 20 feet will be permitted.

!Codealert

A cricket or saddle shall be installed on the ridge side of any chimney or penetration greater than 30 inches wide as measured perpendicular to the slope. Cricket or saddle coverings shall be of sheet metal or of the same material as the roof covering.

- On one-story buildings, combustible unroofed mechanical-equipment screens, fences, or similar enclosures are permitted when they are located at a fire-separation distance of at least 20 feet from adjacent property lines and do not exceed 4 feet in height above the roof surface.
- Dormers must be of the same type of construction as the roof on which they are placed or of the exterior walls of the building.

If you decide to put a tank on a rooftop, be aware of the code requirements. Tanks that have a capacity of more than 500 gallons placed in or on a building must be supported on masonry, reinforced concrete, steel, or Type IV construction provided that, where such supports are located in the building above the lowest story, the support must be fire-resistance-rated as required for Type IA construction.

Every tank must have a pipe fitted with a suitable quick-opening valve for emptying the contents in an emergency. Do not ever place a tank over or near stairs or an elevator shaft unless there is a solid roof or floor underneath. A building rooftop may have a tower, spire, dome, or cupola, as long as they conform to the code for these types of structures. The construction of the projections must match or be no less in fire-resistance construction than for the whole building. Any of these projections that are more than 60 feet high must be entirely constructed of noncombustible materials and must be separated from the building below by construction having a fire-resistance rating of no less than 1.5 hours.

Exterior walls of enclosed towers or spires have the same requirements as the building that they are attached to.

REROOFING

All materials that you use to reroof must be of the same quality as for new construction, except that reroofing does not have to meet the minimum design slope requirements of 1/4 unit vertical in 12 units horizontal for roofs that provide positive roof drainage. Structural roof components must be capable of supporting the roof-covering system and the material and equipment loads of the system. There is a very important provision to remember when reproofing. Never install a new roof without first removing all existing layers of roof covering where any of the following conditions occur:

- Where the existing roof or roof covering is water-soaked or has deteriorated to the point that the existing roof or roof covering is not adequate as a base for additional roofing
- Where the existing roof covering is wood shake, slate, clay, cement, or asbestos-cement tile
- Where the existing roof has two or more applications of any type of roof covering

There are exceptions to these conditions:

• Complete and separate roofing systems, such as standing-seam metal roof systems, that are designed to send the rood loads directly to the building's structural system and that do not rely on existing roofs and roof coverings for support, do not require the removal of existing roof coverings.

• Metal panel, metal shingle, and concrete and clay-tile roof coverings are allowed over existing wood-shake roofs when applied in accordance with this chapter.

• The application of a new protective coating over an existing spray-polyurethane-foam roofing system is permitted without removal of existing roof coverings.

If the application of a new roof covering of wood shingle or shake roof creates a combustible concealed space, you must cover the entire existing surface with gypsum board, mineral fiber, or other approved materials.

There are some instances where you may reinstall materials. This applies only to slate, clay or cement tile that is not cracked, damaged, or broken. You must reconstruct flashings in accordance with approved manufacturer's installation instructions as well.

This concludes Chapter 15.

!Codealert

When reroofing, the application of a new protective coating over an existing spray polyurethane-foam roofing system shall be permitted without tear-off of existing roof coverings.

CHAPTER 16
STRUCTURAL DESIGN

I am going to begin this chapter with definitions, since these terms have changed since the last code.

In this chapter you will also come across letters and notations:

- D = Dead load

- E = Combined effect of horizontal and vertical earthquake-induced forces as defined in Section 12.4.3 of ASCE 7

- E_m = Maximum seismic -oad effect of horizontal and vertical seismic forces as set forth in Section 12.4.3 of ASCE 7

- F = Loads due to fluids with well-defined pressures and maximum heights

- F_a = Flood load

- H = Load due to lateral earth pressure, ground-water pressure, or pressure of bulk materials

- L = Live load, excluding roof live load, including any permitted live-load reduction

- L_r = Roof live load including any permitted live-load reduction

- R = Rain load

- S = Snow load

- T = Self-straining force arising from contraction or expansion resulting from temperature change, shrinkage, moisture change, creep in component materials, movement due to differential settlement, or combinations thereof

- W = Load due to wind pressure

!Definitionalert

Diaphragm, Flexible: A diaphragm flexible for the purpose of distribution of story shear and torsional moment where so indicated in Section 12.3.1 of ASCE 7 and as modified in this chapter of the code.

Fabric Partitions: A partition consisting of a finished surface made of fabric, without a continuous rigid backing, that is directly attached to a framing system in which the vertical framing members are spaced greater than 4 feet on center.

Occupancy Category: A category used to determine structural requirements based on occupancy.

Vehicle Barrier System: A system of building components near open sides of a garage floor, ramp or building wall, that acts as a restraint for vehicles.

CONSTRUCTION DOCUMENTS

Are you aware of what information needs to be included in construction documents? Much important information cannot be left out, such as the size, section, and relative locations of structural members, with floor levels, column centers, and offsets dimensioned. The design loads and other information pertinent to the structural design are described in this section and must be indicated on the construction documents. Construction documents for buildings constructed in accordance with the conventional light-frame-construction provision of Chapter 22 are an exception and must indicate the following structural-design information:

• Floor and roof live loads

• Ground snow lead

• Basic wind speed (3-second gust) in miles per hour and wind exposure

• Seismic-design category and site class

• Flood design data if located in flood-hazard area

The floor live load that has been uniformly distributed and used in the design must be indicated. If a reduction is allowed, that, too, must be indicated. Construction documents must include wind design data, regardless of whether wind loads govern the design of the lateral-force-resisting system of the building.

When wind enters a building, internal pressure and external suction create a combined load, causing the building in the worst case to crumble into a pile of sticks. The following list contains wind data to include:

- Basic wind speed in miles per hour
- Wind-importance factor and occupancy category
- Wind exposure; if more than one wind exposure is used, the exposure and applicable wind direction must be indicated
- Applicable internal-pressure coefficient
- Components and cladding

Earthquake design data is must be included. Starting from the seismic importance and occupancy factor to the analysis procedure used, this important component must be included in any construction documents. If your building is located in a flood-hazard area, the documentation pertaining to design, if required, and the following information, referenced to the data on the community's Flood Insurance Rate Map or FIRM, must be included, regardless of whether flood loads govern the design of the building:

- In flood-hazard areas not subject to high-velocity wave action, the elevation of the proposed lowest floor, including the basement; and the elevation to which any nonresidential building will be dry-flood-proofed
- Elevations to which any nonresidential building will be dry-flood-proofed
- The proposed elevation of the bottom of the lowest horizontal structural member of the lowest floor, including the basement

You must indicate any special loads that are applicable to the design of the building. Keep in mind that it is illegal to allow a load greater than permitted by the code to be placed upon any floor or roof of a building. Occupancy permits will not be issued until the floor-load signs have been installed.

GENERAL DESIGN REQUIREMENTS

Buildings and structures have to be constructed in accordance with the strength, load and resistance, and allowable stress design requirements given in the applicable material chapters in the code. The structural components of any building must be designed and constructed to safely support the factored loads in load combinations that have been designed according to the code. You must be careful not to exceed the appropriated strength-limit states for the construction materials. You must consult the building official for loads and forces for occupancies or uses that are not covered in the chapter. All structural systems and members must be designed to limit deflections and lateral drift. Deflections of any structural members cannot be more than allowed in Table 16.1.

Account equilibrium, general stability, and geometric compatibility must be taken into account to determine the load effects on structural members and connections. Residual deformations, which tend to accumulate on members under repeated service loads, must be included in the analysis in regard to added eccentric-

TABLE 16.1 Deflection limits. [a,b,c,h,i]

CONSTRUCTION	L	S or W^f	$D + L^{d,g}$
Roof members:[c]			
Supporting plaster ceiling	$l/360$	$l/360$	$l/240$
Supporting nonplaster ceiling	$l/240$	$l/240$	$l/180$
Not supporting ceiling	$l/180$	$l/180$	$l/120$
Floor members	$l/360$		$l/240$
Exterior walls and interior partitions:			
With brittle finishes	—	$l/240$	—
With fexible finishes	—	$l/120$	—
Farm buildings	—	—	$l/180$
Greenhouses	—	—	$l/120$

For SI: 1 foot = 304.8 mm

a. For structural roofing and siding made of formed metal sheets, the total load deflection shall not exceed $l/60$. For secondary roof structural members supporting formed metal roofing, the live load deflection shall not exceed $l/150$. For secondary wall members supporting formed metal siding, the design wind load deflection shall not exceed $l/90$. For roofs, this exception only applies when the metal sheets have no roof covering.

b. Interior partitions not exceeding 6 feet in height and flexible, folding and portable partitions are not governed by the provisions of this section. The deflection criterion for interior partitions is based on the horizontal load defined in Section 1607.13.

c. See Section 2403 for glass supports.

d. For wood structural members having a moisture content of less than 16 percent at time of installation and used under dry conditions,t he deflection resulting from $L + 0.5D$ is permitted to be substituted for the deflection resulting from $L + D$.

e. The above deflection do not ensure against ponging. Roofs that do not have sufficient slope or camber to assure adequate drainage shall be investigated for ponging. See Section 1611 for rain and ponging requirements and Section 1503.4 for roof drainage requirements.

f. The wind load is permitted to be taken as 0.7 times the "component and cladding" loads for the purpose of determining deflection limits herein.

g. For steel structural members, the dead load shall be taken as zero.

h. For aluminum structural members or aluminum panels used in skylights and sloped glazing framing, roofs or walls of sunroom additions or patio covers, not supporting edge of glass or aluminum sandwich panels, the total load deflection shall not exceed $1/60$. For aluminum sandwich panels used in roofs or walls of sunroom additions or patio covers, the total load deflection shall not exceed $1/120$.

i. For cantilever members, l shall be taken as twice the length of the cantilever.

ities expected to occur during their service life. Any system or method of construction to be used must be based on a rational analysis. This analysis must result in a system that provides a complete load path capable of transferring loads from the origin point to the load-resisting elements.

Any rigid elements that are assumed not to be a part of the lateral-force-resisting system are allowed to be incorporated into buildings so long as their effect on the action of the system is considered and provided for in the design. In instances where diaphragms are flexible or are allowed to be analyzed as flexible, provisions will be made for the increased forces induced on resisting elements of the structural system resulting from torsion due to eccentricity between the center of application of the lateral forces and the center of rigidity of the lateral-force-resisting system.

Table 16.2 assigns occupancy categories of buildings and other structures. For instance, buildings and structures that have a relatively low hazard for people if there is an emergency issue are categorized as occupancy I. As you can see, this table doesn't give any information for buildings that have two or more occupancies. In such cases, the structure must be assigned to the most restrictive classification.

For example, if one part of the building is a storage facility and the other is a jail or detention center, the building must be categorized as occupancy III, because there are people who occupy the jail. If the building or structure has two separate portions with separate entrances, they must be categorized separately. The building official is authorized to require an engineering analysis, a load test, or even both for any building when there is a question of safety regarding the construction of the occupancy in question.

To resist the uplift and sliding forces that result from the use of prescribed loads, you must anchor the roof to walls and columns and, in turn, anchor the walls and columns to foundations. If your building has concrete and masonry walls, these must be anchored to floors, roofs, and other structural elements that are used to provide lateral support for the wall. Realize that the required anchors that you use in masonry walls of hollow units or cavity walls must be embedded in a reinforced, grouted structural element of the wall.

Structural elements, components, and cladding must be designed to resist forces from earthquakes and wind with consideration to overturning, sliding, and uplift. As a contractor you must provide continuous load paths for transmitting these forces to the foundation.

LOAD COMBINATIONS

In cases where allowable-stress or working-stress design is permitted by the code, structures and portions thereof must resist the most critical effects resulting from the following combinations of loads:

$$1.4\,(D + F) \qquad\qquad \text{(Equation 16.1)}$$

$$1.2\,(D + F + T) + 1.6\,(L + H) + 0.5\,(L_r \text{ or } S \text{ or } R) \qquad\qquad \text{(Equation 16.2)}$$

$$1.2\,D + 1.6\,(L_r \text{ or } S \text{ or } R) + (f_1 L \text{ or } 0.8W) \qquad\qquad \text{(Equation 16.3)}$$

$$1.2\,D + 1.6W + f_1 L + 0.5\,(L_r \text{ or } S \text{ or } R) \qquad\qquad \text{(Equation 16.4)}$$

$$1.2\,D + 1.0E + f_1 L + f_2 S \qquad\qquad \text{(Equation 16.5)}$$

TABLE 16.2 Occupancy category of buildings and other structures.

OCCUPANCY CATEGORY	NATURE OF OCCUPANCY
I	Buildings and other structures that represent a low hazard to human life in the event of a failure, including but not limited to: • Agricultural facilities • Certain temporary facilities • Minor storage facilities
II	Buildings and other structures except those listed in Occupancy Categories I, III and IV
III	Buildings and other structures that represent a substantial hazard to human life in the event of a failure, including but not limited to: • Covered structures whose primary occupancy is public assembly with an occupant load greater than 300 • Buildings and other structures with elementary school, secondary school or day care facilities with an occupant load greater than 250 • Buildings and other structures with an occupant load greater than 500 for colleges or adult education facilities • Health care facilities with an occupant load of 50 or more resident patients, but not having surgery or emergency treatment facilities • Jails and detention facilities • Any other occupancy with an occupant load greater than 5,000 • Power-generating stations, water treatment for potable water, waste water treatment facilities and other public utility facilities not included in Occupancy Category IV. • Buildings and other structures not included in Occupancy Category IV containing sufficient quantities of toxic or explosive substances to be dangerous to the public if released.
IV	Buildings and other structures designated as essential facilities, including but not limited to: • Hospitals and other health care facilities having surgery or emergency treatment facilities • Fire, rescue and police stations and emergency vehicle garages • Designated earthquake, hurricane or other emergency shelters • Designated emergency preparedness, communication, and operation centers and other facilities required for emergency response • Power-generating stations and other public utility facilities required as emergency backup facilities for Occupancy Category IV structures • Structures containing highly toxic materials as defined by Section 307 where the quantity of the material exceeds the maximum allowable quantities of Table 307.1(2) • Aviation control towers, air traffic control centers and emergency aircraft hangers • Buildings and other structures having critical national defense functions • Water treatment facilities required to maintain water pressure for fire suppression

!Definitionalert

Dead Load: The weight of materials of construction incorporated into the building, including but not limited to walls, floors, roofs, ceiling, stairways, and other similarly incorporated architectural and structural items, and the weight of fixed service equipment, such as cranes, plumbing stacks and risers, electrical feeders, and heating, ventilating, and air-conditioning systems.

$0.9D + 1.6W + 1.6H$ (Equation 16.6)

$0.9D + 1.0E + 1.6H$ (Equation 16.7)

f_1 = 1 for floors in places of public assembly, for live loads in excess of 100 pounds per square foot (4.79 kN/m^2), and for parking garage live load, and

   = 0.5 for other live loads

f_2 = 0.7 for roof configurations (such as saw tooth) that do not shed snow off the structure, and

   = 0.2 for other roof configurations

Please note these exceptions:

- Crane-hook loads do not need to be combined with the roof live load or with more than three-fourths of the snow load or one-half of the wind load; flat roof loads of 30 psf or less do not need to be combined with seismic loads unless the flat-roof snow load exceed 30 psf, in which case 20 percent must be combined with seismic loads.

- Keep in mind that increases in allowable stresses specified in the referenced standards cannot be used with the load combination that I've listed above, except that a duration of load increase is permitted in accordance with Chapter 23.

- There may be times when you will be allowed to use alternative basic load combinations that include wind or seismic loads. In these cases, allowable stresses can be increased or load combinations reduced where permitted by the materials chapters of this book or the referenced standards.

- Be confident that you can make these changes by reading chapter 23 and understanding the referenced standards carefully.

Heliport and helistop landing areas are occupancies where the load factors are of a serious nature and landing areas must be designed in accordance with the following:

> **!Code**alert
>
> A fabric partition consisting of a finished surface made of fabric, without a continuous rigid backing, that is directly attached to a framing system in which the vertical framing members are spaced greater than 4 feet on center.

- Dead load, D, plus the gross weight of the helicopter, D_h, plus snow load, S
- Dead load, D, plus two single concentrated-impact loads, L, approximately 8 feet apart applied anywhere on the landing area, with a magnitude of 0.75 times the gross weight of the helicopter
- Both loads acting together total one-and-one-half times the gross weight of the helicopter
- Dead load, D, plus a uniform live load, L, of 100 psf

Please note that there is an exception to this provision, which is also a change in the code, as follows:

- Landing areas designed for helicopters with gross weights not exceeding 3,000 pounds in accordance with points 1 and 2 (above list) must be allowed to be designed using a 40 psf uniform live load in point 3 (above list), provided the landing area is identified with a 3,000-pound weight limitation
- The 40 psf uniform live load cannot be reduced and the landing-area weight limitation must be indicated by the numeral 3
- The appropriate location is in the bottom-right corner of the landing area as viewed from the primary approach
- The landing-area weight-limitation designation must be a minimum of 5 feet in height

DEAD LOADS

Dead loads are the weights of various structural members and objects that are permanently attached to the structure. There are two types of dead loads: building dead loads and collateral dead loads. A building dead load is the actual building system, such as a roof or floor and the materials used for covering such as decking, felt, and hinges. Collateral dead loads are the weight of the permanent materials, such as drywall, sprinklers, and electrical systems. Collateral dead loads do not include the weight of the actual building system. In the absence of definite information, values will be subject to the approval of the building official. We call these "dead loads" because they are unable to be moved.

LIVE LOADS

One can't have a dead load without a live load, so let's explore this subject a bit more. Live loads are not permanent and can change in magnitude. Live loads include items that can be found inside a building, such as furniture, safes, people, or stored materials. Environmental effects, such as earthquakes, wind, and snow, which have the power to change and cause potential damage or failure to a building are also considered live loads. However, the live load must be determined in accordance with a method approved by the building official.

Table 16.3 shows the minimum uniformly distributed live loads and minimum concentrated live loads allowed by the code. In no case will the maximum load be less than the minimum unit load as required by this table.

Concentrated loads, forces localized over a relatively small area such as floors and similar surfaces, must be designed to support live loads that are uniformly distributed. In office buildings and in other buildings where partition locations are subject to change, requirements for partition weight must be followed. Whether these partitions are shown on the construction documents or not, you must allow for them, unless the specified live load is more than 80 psf. In case you're wondering, the partition load cannot be less than a uniformly distributed live load of 15 psf.

Truck and bus garages are among the many buildings that require a minimum live load. Both uniform and concentration loads must be uniformly distributed over a 10-foot-wide ad placed within their individual lanes. This will produce the maximum stress in each structural member but not on both at the same time. Keep in mind that all garages accommodating trucks and buses must also be designed with an approved method that contains provisions for traffic railings.

!Definitionalert

Wheel Load: The vertical force without impact produced on a crane wheel bearing on a runway rail or suspended from a runway beam. Maximum wheel load occurs with the crane at rated capacity and the trolley positioned to provide maximum vertical force at one set of wheels.

TABLE 16.3 Minimum uniformly distributed live loads and minimum concentrated live loads.[g]

OCCUPANCY OR USE	UNIFORM (psf)	CONCENTRATED (lbs.)
1. Apartments (see residential)	—	—
2. Access floor systems 　Office use 　Computer use	 50 100	 2,000 2,000
3. Armories and drill rooms	150	—
4. Assembly areas and theaters 　Fixed seats (fastened to floor) 　Follow spot, projections and control rooms 　Lobbies 　Movable seats 　Stages and platforms	 60 50 100 100 125	 —
5. Balconies 　On one- and two-family residences only, 　and not exceeding 100 sq ft	100 60	—
6. Bowling alleys	75	
7. Catwalks	40	
8. Dance halls and ballrooms	100	—
9. Decks	Same as occupancy served [h]	—
10. Dining rooms and restaurants	100	—
11. Dwellings (see residential)	—	—
12. Cornices	60	—
13. Corridors, except as otherwise indicated	100	—
14. Elevator machine room grating (on area of 4 in^2)	—	
15. Finish light floor plate construction 　(on area of 1 in^2)	—	
16. Fire escapes 　(on single family dwellings only)	100 40	—
17. Garages (passenger vehicles only) 　Trucks and buses	40 See Section 1607.6	Note a
18. Grandstands (see stadium and arena bleachers)	—	—
19. Elevator machine room grating (on area of 4 in^2)	100	—
20. Handrails, guards and grab bars	See Section 1607.7	
21. Hospitals 　Corridors above first floor 　Operating rooms, laboratories 　Patient rooms	 80 60 40	 1,000 1,000 1,000
22. Hotels (see residential)	—	—

TABLE 16.3 Minimum uniformly distributed live loads and minimum concentrated live loads.[g] *(continued)*

OCCUPANCY OR USE	UNIFORM (psf)	CONCENTRATED (lbs.)
23. Libraries		
Corridors above first floor	80	1,000
Reading rooms	60	1,000
Stack rooms	150[b]	1,000
24. Manufacturing		
Heavy	250	3,000
Light	125	2,000
25. Marquees	75	—
26. Office buildings		
Corridors above first floor	80	2,000
File and computer rooms shall be designed for heavier loads based on anticipated occupancy	—	—
Lobbies and first-floor corridors	100	2,000
Offices	50	2,000
27. Penal institutions		
Cell blocks	40	—
Corridors	100	
28. Residential		
One- and two-family dwellings		
Uninhabitable attics without storage[i]	10	
Uninhabitable attics with limited storage[i,j,k]	20	
Habitable attics and sleeping areas	30	—
All other areas except balconies and decks	40	
Hotels and multiple-family dwellings		
Private rooms and corridors serving them	40	
Public rooms and corridors serving them	100	
29. Reviewing stands, grandstands and bleachers	Note c	
30. Roofs		
All roof surfaces subject to maintenance workers		300
Awnings and canopies		
Fabric construction supported by a light-weight rigid skeleton structure	5 nonreduceable	
All other construction	20	
Ordinary flat, pitched, and curvd roofs	20	
Primary roof members, exposed to a work floor		
Single panel point of lower chord of roof trusses or any point along primary structural members supporting roofs		
Over manufacturing, storage warehouses, and repair garages		2,000
All other occupancies		300
Roofs used for other special purposes	Note 1	Note 1
Roofs used for promenade purposes	60	
Roofs used for roof gardens or assembly purposes	100	

TABLE 16.3 Minimum uniformly distributed live loads and minimum concentrated live loads.[g] *(continued)*

OCCUPANCY OR USE	UNIFORM (psf)	CONCENTRATED (lbs.)

31. Schools
   Classrooms
   Corridors above first floor
   First-floor corridors
	
	40
	80
	100
		
		1,000
		1,000
		1,000

32. Scuttles, skylight ribs and accessible ceilings	—	200
33. Sidewalks, vehicular driveways and yards, subject to trucking	250[d]	8,000[e]
34. Skating Rinks	100	—

35. Stadiums and arenas
   Bleachers
   Fixed seats (fastened to floor)
	
	100[c]
	60[c]
		
		—

36. Stairs and exits
   One- and two-family dwellings
   All other
	
	40
	100
		
		Note f

37. Storage warehouses (shall be designed for heavier loads if required for anticipated storage)
   Heavy
   Light
	
	250
	125	

38. Stores
   Retail
      First floor
      Upper floors
   Wholesale, all floors
	
	100
	75
	125
		
		1,000
		1,000
		1,000

39. Vehicle barriers	See Section 1607.7.3	
40. Walkways and elevated platforms (other than exitways)	60	—
41. Yards and terraces, pedestrians	100	—

For SI: 1 inch = 25.4 mm, 1 square inch = 645.16 mm², 1 square foot = 0.0929 m²,
    1 pound per square foot = 0.0479 kN/m², 1 pound = 0.004448 kN, 1 pound per cubic foot = 16 kg/m³

a. Floors in garages or portions of buildings used for the storage of motor vehicles shall be designed for the uniformly distributed live loads of Table 1607.1 or the following concentrated loads: (1) for garages restricted to vehicles accommodating not more than nine passengers, 3,000 pounds acting on an area of 4.5 inches by 4.5 inches; (2) for mechanical parking structures without slab or deck which are used for storing passenger vehicles only 2,250 pounds per wheel.

b. The loading applies to stack room floors that support nonmobile, double-faced library bookstacks, subject to the following limitations:
   1. The nominal bookstack unit height shall not exceed 90 inches;
   2. The nominal shelf depth shall not exceed 12 inches for each face; and
   3. Parallel rows of double-faced bookstacks shall be separated by aisles not less than 36 inches wide.

c. Design in accordance with the ICC *Standard on Bleachers, Folding and Telescopic Seating and Grandstands.*

d. Other uniform loads in accordance with an approved method which contains provisions for truck loading shall also be considered where appropriate.

e. The concentrated wheel load shall be applied on an area of 20 square inches.

f. Minimum concentrated load on stair treads (on area of 4 square inches) is 300 pounds.

If you take a look at Table 16.3, you will see that residential apartments have requirements for loads. Handrail assemblies and guards have to be able to resist 200 pounds in a single concentrated load in any direction at any point along the top and must have attachment devices and supporting structure that can transfer the loading.

With the exception of roof uniform live loads, all other minimum uniformly distributed live loads in Table 16.3 are allowed to be reduced in accordance with certain parts of this section. It is up to you to read and understand these exceptions if they pertain to your building or structure. Note that live loads over 100 psf cannot be reduced unless the following are true:

• The live loads for members supporting two or more floors are permitted to be reduced by a maximum of 20 percent, but the live load cannot be less than the reduced design live load per square foot.

• For uses other than storage, where approved, additional live-load reductions are permitted where shown by a registered design professional that a rational approach has been used and that the reductions are necessary.

As an alternative to the section above, floor live loads are permitted to be reduced in accordance with the following provisions. These reductions apply to slab systems, beams, girders, columns, piers, walls and foundations:

• A reduction is not permitted in Group A occupancies.

• A reduction is not permitted where the live load exceeds 100 psf except that the design live load for members supporting two or more floors is permitted to be reduced by 20 percent.

• A reduction is not permitted in passenger parking garages except that the live loads for members supporting two or more floors may be reduced by a maximum of 20 percent.

• For all live loads that are not more than 100 psf, the design live load for any structural member supporting 150 square feet or more is allowed to be reduced in accordance with the following equation:

$D + F$ (Equation 16.8)

Construction crews must consider not only how loading conditions might affect a structure but also how loads are distributed. This affects both floor and roof loads. Where the uniform floor live load is involved with the design of structural members, the minimum applied loads must be the full dead loads on all spans in combination with the floor live loads in order to create the greatest weight-bearing effect at each location under consideration.

Where uniform roof loads are reduced to less than 20 psf and are involved in the design of structural members, the minimum applied loads must also be the full dead load on adjacent spans or on alternate spans, whichever produces the greatest effect.

If you look back at Table 6.3, you will find that the minimum uniformly distributed roof live loads are allowed to be reduced but only as follows: ordinary flat, pitched, and curved roofs are allowed to be designed for a reduced roof live load as specified in the equation below.

$$D + H + F + L + T \hspace{4cm} \text{(Equation 16.9)}$$

However, you must be careful, because if your workers are using scaffolding and materials during maintenance and repair jobs, a lower roof load than specified cannot be used unless approved by the building official. The safety of humans must always be first on your mind.

Cranes are often used in construction. The crane live load must always be the rated capacity of the crane. Design loads of moving bridge cranes and monorail cranes must include the maximum wheel loads and the vertical-impact, lateral, and longitudinal forces made by the moving crane. The design load includes connections and support brackets as well.

Wheel loads of a crane have a maximum load and can be increased by the percentages, shown below, to determine the induced vertical impact or vibration force:

• Monorail cranes (powered): 25 percent

• Cab-operated or remotely operated bridge cranes (powered): 25 percent

• Pendant-operated bridge cranes (powered): 10 percent

• Bridge or monorail cranes with hand-geared bridge, trolley, and hoist: 0 percent

SNOW AND WIND LOADS

Other loads include elemental or weather-related loads such as snow and wind. Design snow loads are determined in accordance with Chapter 7 of ASCE 7 or Figure 16.1 for the neighboring United States. Please see Table 16.4 for the ground snow loads for Alaskan locations. You must use extreme-value statistical analysis to determine ground snow loads that are found in the vicinity of the site using a value with a 2-percent annual probability of being exceeded. Chapter 6 of ASCE 7 is used to determine wind loads on every building or structure.

!Definitionalert

Site Class: A classification assigned to a site based on the types of soils present and their engineering properties as defined in this chapter of the code.

FIGURE 16.1 Ground snow loads, p_g, for the United States (psf).

The type of opening protection, the basic wind speed, and the exposure category for a site is determined in this chapter of the code. Wind can be assumed to come from any horizontal direction, and wind pressures can be assumed to behave normally at the surface that is being considered.

There are several areas that are defined as hurricane-prone regions. These regions include the U.S. Atlantic Ocean and Gulf of Mexico coasts, where the basic wind speed is greater than 90 mph, and Hawaii, Puerto Rico, Guam, Virgin Islands, and American Samoa. Portions of these areas that are within 1 mile of the coastal mean-high-water line, where the basic wind speed is 110 mph or greater,

FIGURE 16.1 Ground snow loads, p_g, for the United States (psf). *(continued)*

portions of hurricane-prone regions where the basic wind speed is 120 mph or greater, and Hawaii are also considered wind-borne-debris regions.

For basic wind speeds, in mph, the wind loads are determined by Figure 16.2. Find your local jurisdiction requirements to determine basic wind speed for the special wind regions indicated near mountainous terrain and gorges. The basic wind speed can be converted using Table 16.5.

For each wind direction considered, an exposure category that adequately reflects the characteristics of ground-surface irregularities must be determined for

TABLE 16.4 Occupancy category of buildings and other structures.

LOCATION	POUNDS PER SQUARE FOOT	LOCATION	POUNDS PER SQUARE FOOT	LOCATION	POUNDS PER SQUARE FOOT
Adak	30	Galena	60	Petersburg	150
Anchorage	50	Gulkana	70	St. Paul Islands	40
Angoon	70	Homer	40	Seward	50
Barrow	25	Juneau	60	Shemya	25
Barter Island	35	Kenai	70	Sitka	50
Bethol	40	Kodiak	30	Talkeetna	120
Big Delta	50	Kotzebue	60	Unalakleet	50
Cold Bay	25	McGrath	70	Valdez	160
Cordova	100	Nenana	80	Whittier	300
Fairbanks	60	Nome	70	Wrangell	60
Fort Yukon	60	Palmer	50	Yakutat	150

For SI: 1 pound per square foot = 0.0479 kN/m^2.

the site at which the building or structure is to be constructed. You must take into account any variation in ground-surface roughness that arises from natural topography and vegetation as well as constructed features. A ground-surface roughness within each 45-degree sector is determined for a distance upwind of the site as defined from the categories below:

• Surface Roughness B: Urban and suburban areas, wooded areas, or other terrain with numerous closely spaced obstructions having the size of single-family dwellings or larger.

• Surface Roughness C: Open terrain with scattered obstructions having heights generally less than 30 feet; this category includes flat, open country, grasslands, and all water surfaces in hurricane-prone regions.

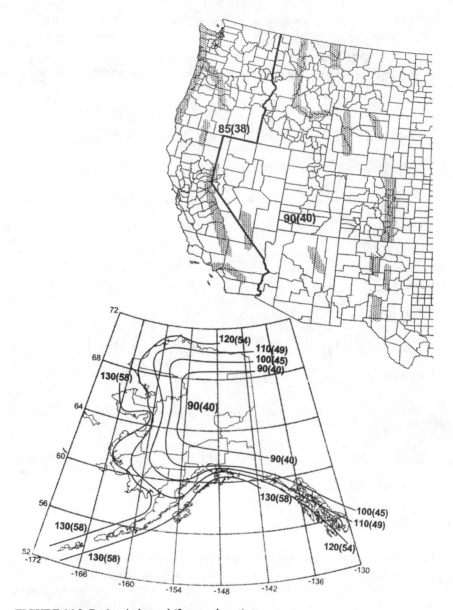

FIGURE 16.2 Basic wind speed (3-second gust).

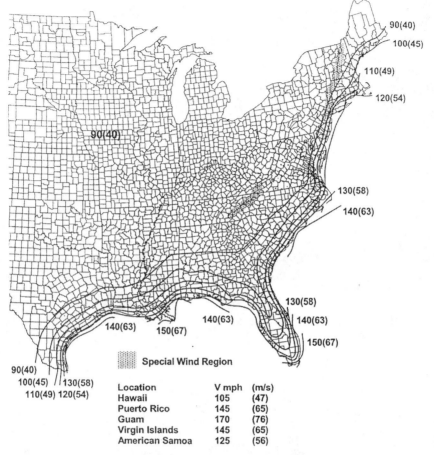

Location	V mph	(m/s)
Hawaii	105	(47)
Puerto Rico	145	(65)
Guam	170	(76)
Virgin Islands	145	(65)
American Samoa	125	(56)

Notes:
1. Values are nominal design 3-second gust wind speeds in miles per hour (m/s) at 33 ft (10 m) above ground for Exposure C category.
2. Linear interpolation between wind contours is permitted.
3. Islands and coastal areas outside the last contour shall use the last wind speed contour of the coastal area.
4. Mountainous terrain, gorges, ocean promontories, and special wind regions shall be examined for unusual wind conditions.

FIGURE 16.2 Basic wind speed (3-second gust). *(continued)*

TABLE 16.5 Equivalent basic wind speeds.[a,b,c]

V_{3S}	85	90	100	105	110	120	125	130	140	145	150	160	170
V_{fm}	71	76	85	90	95	104	109	114	123	128	133	142	152

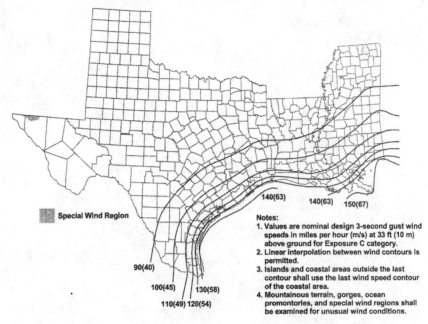

140(63)   140(63)   150(67)

Special Wind Region

Notes:
1. Values are nominal design 3-second gust wind speeds in miles per hour (m/s) at 33 ft (10 m) above ground for Exposure C category.
2. Linear interpolation between wind contours is permitted.
3. Islands and coastal areas outside the last contour shall use the last wind speed contour of the coastal area.
4. Mountainous terrain, gorges, ocean promontories, and special wind regions shall be examined for unusual wind conditions.

90(40)

100(45)   130(58)

110(49) 120(54)

FIGURE 16.2 Basic wind speed (3-second gust), western Gulf of Mexico hurricane coastline. *(continued)*

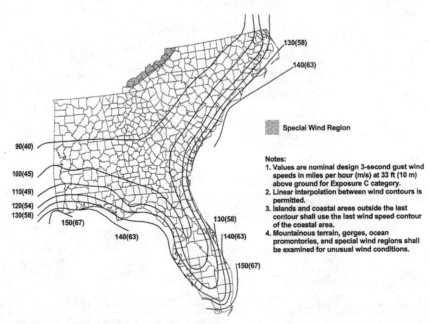

130(58)

140(63)

Special Wind Region

Notes:
1. Values are nominal design 3-second gust wind speeds in miles per hour (m/s) at 33 ft (10 m) above ground for Exposure C category.
2. Linear interpolation between wind contours is permitted.
3. Islands and coastal areas outside the last contour shall use the last wind speed contour of the coastal area.
4. Mountainous terrain, gorges, ocean promontories, and special wind regions shall be examined for unusual wind conditions.

90(40)

100(45)

110(49)

120(54)

130(58)

150(67)

140(63)

130(58)

140(63)

150(67)

Figure 16.2 Basic wind speed (3-second gust), eastern Gulf of Mexico and southeastern U.S. hurricane coastline. *(continued)*

> # !Codealert
> Occupancy category is a category used to determine structural require-ments.

- Surface Roughness D: Flat, unobstructed areas and water surfaces outside hurri-cane-prone regions; this category includes smooth mud flats, salt flats, and un-broken ice.

Roof systems, such as roof decks, are affected by wind loads and must be de-signed to withstand the wind pressures determined in accordance with ASCE 7. If you are like most people, you've applied or intend to apply asphalt shingles to your roof deck for design purposes. Make sure that any asphalt shingles that you are using have been tested to determine the resistance of the sealant to uplift forces using ASTM D 6381. You can find yourself in a bit of a mess if you disregard this. All it takes is a heavy wind to blow those shingles right off the roof of your build-ing or structure. If you've decided on rigid tile for your roof covering, make sure you note that the code requirement is not the same. Look at Equation 16.10 for fur-ther explanation:

$$D + H + F + (L_r \text{ or } S \text{ or } R) \qquad \text{(Equation 16.10)}$$

Concrete and clay roof tiles complying with the following limitations must be designed to withstand the aerodynamic uplift moment:

- The roof tiles must be loose-laid on battens, mechanically fastened, or mortar/adhesive-set
- The roof tiles must be installed on solid sheathing that has been designed as com-ponents and cladding.
- An underlayment must be installed in accordance with Chapter 15.
- The tile must be single-lapped interlocking with a minimum head lap of no less than 2 inches.
- The length of the tile must be between 1.0 and 1.75 feet.
- The maximum thickness of the tail of the tile cannot be more than 1.3 inches.
- Roof tiles using mortar-set or adhesive-set systems must have at least two-thirds of the tile's area free of mortar or adhesive contact.

!Codealert

A system of building components near open sides of a garage floor or ramp or building walls that act as restraints for vehicles is a vehicle barrier system.

SOIL LATERAL LOADS

This section refers to basements, foundations, and retaining walls that have to be designed to resist lateral soil loads. Soil loads specified in Table 16.6 must be used as the minimum design lateral soil loads unless specified otherwise in a soil-investigation report. Make sure that you have the investigation report approved by your building official. Basement walls with restricted horizontal movement have to be designed for at-rest pressure. Note that basement walls that do not extend more than 8 feet below grade and supporting flexible floor systems may be designed for active pressure.

RAIN LOADS

When designing a building or structure, engineers and architects must consider the loads, both external and internal, that a building must endure. The building is then designed to resist these loads. A type of external load is rain. When designing a building roof, the goal is to have each portion of the roof sustain the load of rainwater that will accumulate on it if the primary drainage system is not working properly. This is also true for uniform loads caused by water that has risen above the inlet of the secondary drainage system.

A change in the code has a provision for roofs equipped with a slope that is less than 1/4 inch per foot. The design calculations must include verification of adequate stiffness to preclude progressive deflection in accordance with Section 8.4 of ASCE 7. One drainage system is simply not enough.

Roofs that are equipped with hardware to control the rate of drainage must be equipped with a secondary drainage system. All secondary drainage systems must be at a higher elevation that limits any accumulation of water on the roof above that elevation.

TABLE 16.6 Soil lateral load.

DESCRIPTION OF BACKFILL MATERIAL	UNIFIED SOIL CLASSIFICATION	DESIGN LATERAL SOIL LOAD[a] (pound per square foot per foot of depth)	
		Active Pressure	At-rest Pressure
Well-graded, clean gravels; gravel-sand mixes	GW	30	60
Poorly graded clean gravels; gravel-sand mixes	GP	30	60
Silty gravels, poorly graded gravel-sand mixes	GM	40	60
Clayey gravels, poorly graded gravel-and-clay mixes	GC	45	60
Well-graded, clean sands; gravelly sand mixes	SW	30	60
Poorly graded clean sands; sand-gravel mixes	SP	30	60
Silty sands, poorly graded sand-silt mixes	SM	45	60
Sand-silt clay mix with plastic fines	SM-SC	45	100
Clayey sands, poorly graded sand-clay mixes	SC	60	100
Inorganic silts and clayey silts	ML	45	100
Mixture of inorganic silt and clay	MLCL	60	100
Inorganic clays of low to medium plasticity	CL	60	100
Organic silts and silt clays, low plasticity	OL	Note[b]	Note[b]
Inorganic clayey silts, elastic silts	MH	Note[b]	Note[b]
Inorganic clays of high plasticity	CH	Note[b]	Note[b]
Organic clays and silty clays	OH	Note[b]	Note[b]

For SI: 1 pound per square foot per foot of depth = 0.157 kPa/m, 1 foot = 304.8 mm.

a. Design lateral soil loads are given for moist conditions for the specified soils at their optimum densities. Actual field conditions shall govern. Submerged or saturated soil pressures shall include the weight of the buoyant soil plus the hydrostatic loads.

b. Unsuitable as backfill material.

c. The definition and classification of soil materials shall be in accordance with ASTM D 2487.

FLOOD LOADS

Flooding occurs when excess water overflows from bodies of water onto adjacent land. Flood areas are established by local government officials and shown in a flood-hazard map and other supporting data. This map has to include any areas with special flood hazards that have been identified by the Federal Emergency Management Agency and presented as an engineering report. This report has to be written in the following manner: "The Flood Insurance Agency for {NAME OF JURISDICTION}", dated {INSERT DATE OF ISSUANCE}, as amended." Be sure to include the Flood Insurance Rate Map (FIRM), the Flood Boundary and Flooding Map (FBFM), and any supporting data relating to the flood area.

Floods kill people and destroy homes in parts of the world every year. Obviously, any structure or building built in floor-prone areas is permanently at risk. Design and construction of buildings located in flood hazard areas, including those areas that are at risk for high-velocity wave action, must be in accordance with ASCE 24. See the following lists for documentation that must be prepared and sealed by a registered design professional and submitted to the building official. The first list is for construction in flood-hazard areas not to subject high-velocity wave action:

- The elevation of the lowest floor, including the basement, as required by the lowest floor elevation inspection in Chapter 1, must be shown.

- For fully enclosed areas below the design flood elevation where provisions to allow for the automatic entry and exit of floodwaters do not meet the minimum requirements in Section 2.6.2.1 of ASCE 24, construction documents must include a statement that the design will provide for equalization of hydrostatic flood forces in accordance with Section 2.6.2.2 of ASCE 24.

- For dry-flood-proofed nonresidential buildings, construction documents must include a statement that the dry flood-proofing is designed in accordance with ASCE 24.

For construction in flood-hazard areas subject to high-velocity wave action, include the following:

- The elevation of the bottom of the lowest horizontal structural member as required by the lowest floor elevation inspection in Chapter 1 must be shown.

- Construction documents must include a statement that the building is designed in accordance with ASCE 24, including that the pile or column foundation and building or structure to be attached is designed to be anchored to resist flotation, collapse, and lateral movement due to the effect of wind and flood loads acting simultaneously on all building components and other load requirements of Chapter 16.

- For breakaway walls designed to resist a nominal load of less than 10 psf or more than 20 psf, construction documents must include a statement that the breakaway wall is designed in accordance with ASCE 24.

!Codealert

This chapter contains many code changes, additions, and updates. Consult your code book carefully to become aware of all the new code language.

EARTHQUAKE LOADS

The 2006 International Building Code contains a provision that every structure and portion thereof, including nonstructural components that are permanently attached to structures and their supports and attachments, must be designed and constructed to resist the effects of earthquake motions in accordance with ASCE 7. (This does not apply to Chapter 14 on exterior walls or Appendix 11A of the 2006 International Building Code.) You may use this chapter or ASCE 7 to determine the seismic-design category for your structure. Please note the following list of exceptions:

• Detached one- and two-family dwellings that are categorized as A, B, or C are excluded.

• The seismic-force-resisting system of wood-frame buildings that conform to the provisions of Chapter 23 are not required to be analyzed as specified in this section.

• Agricultural-storage structures intended only for incidental human occupancy.

• Structures that require special consideration of their response characteristics and environment and are not addressed by this code or ASCE 7 and for which other regulations provide seismic criteria, such as vehicular bridges, electrical-transmission towers, hydraulic structures, buried utility lines and their accessories, and nuclear reactors, are excluded.

Any existing buildings that you plan on adding to, altering, or modifying must be constructed in accordance with Chapter 34 on existing structures. Please note that this applies to any changes in occupancy as well.

Chapter 17 on special inspections specifies that seismic requirements include a statement of special inspections and must identify the designated seismic systems, seismic-force-resisting systems, and any additional special inspections and testing, such as the applicable standards referenced by the code.

If you take a look at Table 16.7, you will see site-class definitions. This is the classification assigned to a site based on the types of soils present and their engineering properties. Based on site soil properties, the site is classified as Site Class A, B, C, D, E, or F. But what if you're not sure of the soil properties? If you are unsure of the soil properties of your building or structure site and you do not have sufficient detail to determine such information, use the classification Site Class D until the building official or geotechnical data determines which site class the soil belongs in.

TABLE 16.7 Site class definitions.

SITE CLASS	SOIL PROFILE NAME	AVERAGE PROPERTIES IN TOP 100 feet, SEE SECTION 1613.5.5		
		Soil shear wave velocity, $\bar{v}_s$, (ft/s)	Standard penetration resistance, $\bar{N}$	Soil undrained shear strength, $\bar{s}_u$, (psf)
A	Hard rock	$\bar{v}_s > 5{,}000$	N/A	N/A
B	Rock	$2{,}500 < \bar{v}_s \leq 5{,}000$	N/A	N/A
C	Very dense soil and soft rock	$1{,}200 < \bar{v}_s \leq 2{,}500$	$\bar{N} > 50$	$\bar{s}_u \geq 2{,}000$
D	Stiff soil profile	$600 \leq \bar{v}_s \leq 1{,}200$	$15 \leq \bar{N} \leq 50$	$1{,}000 \leq \bar{s}_u \leq 2{,}000$
E	Soft soil profile	$\bar{v}_s < 600$	$\bar{N} < 15$	$\bar{s}_u < 1{,}000$
E	—	Any profile with more than 10 feet of soil having the following characteristics: 1. Plasticity index $PI > 20$, 2. Moisture content $w \geq 40\%$, and 3. Undrained shear strength $\bar{s}_u < 500$ psf		
F	—	Any profile containing soils having one or more of the following characteristics: 1. Soils vulnerable to potential failure or collapse under seismic loading such as liquefiable soils, quick and highly sensitive clays, collapsible weakly cemented soils. 2. Peats and/or highly organic clays ($H > 10$ feet of peat and/or highly organic clay where H = thickness of soil) 3. Very high plasticity clays ($H > 25$ feet with plasticity index $PI > 75$) 4. Very thick soft/medium stiff clays ($H > 120$ feet)		

For SI: 1 foot = 304.8 mm, 1 square foot = 0.0929 m², 1 pound per square foot = 0.0479kPa. N/A = Not applicable.

CHAPTER 17
STRUCTURAL TESTS AND SPECIAL INSPECTIONS

This chapter codifies requirements for inspection and structural tests of buildings and other structures. You will find in this chapter the provisions for quality, workmanship, and materials. All materials of construction must conform to the applicable standards in the code. Any materials, equipment, appliances, systems, or methods of construction that you are planning on using but that are not mentioned in this code must be tested using the methods in this chapter. Testing is necessary for any materials that are of questionable for utility as well. Any used materials that you want to recycle must also meet the minimum requirements for new materials.

APPROVALS

After construction is complete but before the building is occupied, it must pass muster with an approved agency. It is the responsibility of the building official to ensure that this is the case. An approved agency must demonstratee to the building official that it meets all of the applicable requirements. When the agency staff inspects your building, they must be objective and competent. Agencies must disclose possible conflicts of interest. Someone who works for the agency and is also related to the building owner cannot maintain objectivity. Or if the agency has a stake in the building or structure, again, objectivity cannot be maintained.

Agencies must have adequate equipment that is periodically maintained so that the tests performed are accurate, and they must employ experienced personnel who have the required education to perform such duties. All personnel must have experience supervising and evaluating tests and/or inspections. As a building owner, it is OK to ask such questions of the agencies. Since it is your building, it is your right to know the qualifications of such agencies. It is important that you feel comfortable with the agency that is inspecting your building.

!Definitionalert

Approved Agency: An established and recognized agency regularly engaged in conducting tests or furnishing inspection services.

After satisfactory completion of any required tests and after the test reports are submitted, an approval in writing will be given for any material, appliance, equipment, system, or method of construction. A record of this approval, along with any conditions or limitations, is to be kept on file in the office of the building official and viewable to public examination when appropriate.

Specific information from test reports conducted by an approved agency must be given to the building official to determine that the materials meet the applicable code requirement. For example, in Chapter 8 on interior finishes, interior wall or ceiling finishes other than textiles must be tested in accordance with NFPA 286. This is the type of specific information that must be given to the building official.

Another situation that you need to be aware of is when the code requires labeled materials, such as fire-door assemblies, which must be vetted by an approved agency. The agency will test a sample of any materials that bear a label and maintain a record of the tests performed on such materials. An agency will periodically perform an inspection of the product or material that must be labeled. The label must include: manufacturer's or distributor's identification, model number, serial number, and the material's performance characteristics. It also must have the approved agency's identification.

SPECIAL INSPECTIONS

There are many types of constructions that demand special inspections. This responsibility falls upon the owner of the building or the registered design professional who is acting as the owner's agent. One or more special inspectors must be hired to provide inspections during construction. I suggest that you be picky when choosing such inspectors, as the building official must also be satisfied with your choice.

Unless your structure is designed and constructed in accordance with the conventional construction provisions of Chapter 23, any person who applies for a permit must also present a statement of special inspection by the registered design professional as a condition for receiving a permit. You may, however, have this statement prepared by any qualifying person as long as the building official approves and so long as construction is not designed by a registered design professional. This provision is a change in the code—one that you should remember.

The special inspector will keep records of all inspections, and a copy will be given to the building official and to the registered design professional in charge. If any discrepancies are found, the contractor will be notified, and errors must be corrected. Be forewarned: if these changes are not made, the inspector will report this to the building official and to the registered design professional in charge.

Steel Buildings

For buildings and structures made of steel elements, see Table 1704.3 in the International Building Code book. Exceptions are as follows:

- Special inspection of the steel-fabrication process is not required where the fabricator does not perform any welding, thermal cutting, or heating operation of any kind. In these cases, the fabricator is required to submit a detailed procedure for material control that demonstrates the fabricator's ability to maintain records of the procedures used.

- Inspections will be made of any work in progress, and an inspection of all welds will be made prior to the completion of the job The special inspector does not need to be present at all times during the welding of the following items, welding procedures, and qualification of welders are verified before starting any work:

  - Single-pass fillet welds not more than 5/16 inch
  - Floor and roof deck welding
  - Welded studs when used for structural diaphragms
  - Welded sheet steel for cold-formed-steel framing members such as studs and joists
  - Welding of stairs and railing systems

!Definitionalert

Special Inspection: Inspection of the materials, installation, fabrication, erection, or placement of components and connections that need special expertise to ensure compliance with approved construction documents and referenced standards.

!Codealert

A designated seismic system is one where architectural, electrical, and mechanical systems and their components require design in accordance with Chapter 13 of ASCE 7 and where the component importance factor is greater than 1 in accordance with Section 13.1.3 or ASCE 7.

Welding inspections must be in compliance with AWS D1. 1. This is the American Welding Society's code for steel. Welding-inspector qualifications must also be in compliance. The inspector is going to be very scrupulous during any inspection and will inspect such parts as nuts, bolts, washers, and paint and their installation. The inspector will look at the preinstallation testing and calibration process, too.

Concrete Construction

Take a look at Table 1704.4 in the International Building Code book for information on concrete construction.

Below is a list of special conditions where inspections are not required of concrete construction:

- Isolated spread concrete footings of buildings three stories or less in height that are fully supported on earth or rock
- Continuous concrete footing supporting walls of buildings three stories or less in height that are fully supporting on earth or rock where:
    - The footings support walls of light-frame construction
    - The footings are designed in accordance with Table 1805.4.2 of the International Building Code 2006
    - The structural design of the footing is based on a specified compressive strength, f c, no greater than 2,500 pounds per square inch (psi), regardless of the compressive strength specified in the construction documents or used in the footing construction
- Nonstructural concrete slabs supported directly on the ground, including prestressed slabs on grade, where the effective prestress in the concrete is less than 150 psi
- On-grade concrete patios, driveways, and sidewalks

Masonry Construction

The code for masonry construction has had a few changes. While you need not memorize them, I do recommend that you read them carefully and have an understanding of the provisions that the 2006 International Building Code requires for this building material.

All masonry construction is to be inspected and evaluated in accordance with this chapter and is dependent on the classification of the building or structure or of the nature of occupancy as defined by the code. There are three exceptions:

• Special inspections are not required for empirically designed masonry, glass-unit masonry, or masonry veneer. (Please see Chapters 14 and 21 for more detailed information. ACI 530/ASCE 5/TMS 402 also contains information regarding this exception.)

• Inspections are also not required for masonry foundation walls that are constructed in accordance with Table 18.2, which can be found in Chapter 18 of the code.

• Masonry fireplaces, heaters, or chimneys installed or constructed in accordance with Chapter 21 are not required to have special inspections.

However, empirically designed masonry, glass-unit masonry, and masonry veneer in Occupancy IV are required to have special inspections, as are engineered masonry in Occupancy I, II, and III, which must comply with Table 1704.5.1 in the International Building Code book, while engineered masonry in Occupancy IV must comply with Table 1704.5.3 in the International Building Code book.

Soils

Even soils require special inspections. Existing site soil conditions, fill placement, and load-bearing requirements are required by this section and are summarized in Table 17.1. All approved soil reports and any documents prepared by the regis-

!Definitionalert

Pile Foundations: Pile foundations consist of concrete, wood, or steel structural elements either driven into the ground or cast in place. Piles are relatively slender in comparison to their length, with lengths exceeding 12 times the least horizontal dimension. Piles derive their load-carrying capacity through skin friction, end bearing, or a combination of both.

TABLE 17.1 Required verification and inspection of soils.

VERIFICATION AND INSPECTION TASK	CONTINUOUS DURING TASK LISTED	PERIODICALLY DURING TASK LISTED
1. Verify materials below footings are adequate to achieve the design bearing capacity.	—	X
2. Verify excavations are extended to proper depth and have reached proper material.	—	X
3. Perform classification and testing of controlled fill materials.	—	X
4. Verify use of proper materials, densities and lift thicknesses during placement and compaction of controlled fill.	X	—
5. Prior to placement of controlled fill, observe subgrade and verify that site has been prepared properly.	—	X

tered design professional in charge will be used to determine compliance. The inspector will determine that you've used proper materials and have followed procedures in accordance with the code. Note: Special inspection is not required during placement of controlled fill that has a depth totaling 12 inches or less.

Pile Foundations

Special inspections must be performed during installation and testing of pile foundations as required by Table 17.2. All approved soil reports and any documents prepared by the registered design professional in charge will be used to determine compliance. This is also true of pier foundations, and Table 17.3 contains the required verification and inspection for these foundations.

Fire Resistance

Fire resistance requires special inspections. This chapter contains many amendments in the provisions for masonry and foundations. I have provided you with the most up-to-date changes along with the latest tables, which contain the new requirements. Are you aware that structural elements that have been sprayed with fire-resistant materials are subject to special instructions? Inspections of such elements are based on the fire-resistance design as designated in the approved construction documents.

TABLE 17.2   Required verification and inspection of pile foundations.

VERIFICATION AND INSPECTION TASK	CONTINUOUS DURING TASK LISTED	PERIODICALLY DURING TASK LISTED
1. Verify that pile materials, sizes and lengths comply with the requirements.	X	—
2. Determine capacities of test piles and conduct additional load tests, as required.	X	—
3. Observe driving operations and maintain complete and accurate records for each pile.	X	—
4. Verify placement locations and plumbness, confirm type and size of hammer, record number of blows per foot of penetration, determine required penetrations to achieve design capacity, record tip and butt elevations and document any pile damage.	X	—
5. For steel piles, perform additional inspections in accordance with Section 1704.3.	—	—
6. For concrete piles and concrete-filled piles, perform additional inspections in accordance with Section 1704.4.	—	—
7. For specialty piles, perform additional inspsectios as determined by the registered design professional in resonsible charge.	—	—
8. For augered uncased piles and caisson piles, perform inspections in accordance with Section 1704.9.	—	—

It is important that you follow the manufacturer's written instructions and that prepared surfaces have been inspected before you spray the fire-resistant material. The packaging of the material must include approved written instructions, which you must follow before you can obtain approval from the special inspector. Be sure to work in a well-vented area before, during, and after application. No one wants to become sick from the inhalation of fumes.

When spraying the fire-resistant material, be sure that you've read the directions carefully and are aware of the required thickness to be applied to any structural elements. The average thickness of the sprayed fire-resistant materials cannot be less than the thickness that is required by the fire-resistant design. Be sure to record the thickness that you've applied. To determine the thickness of the sprayed fire-resistant material applied to floor, roof, and wall assemblies, refer to

TABLE 17.3 Required verification and inspection of pier foundations.

VERIFICATION AND INSPECTION TASK	CONTINUOUS DURING TASK LISTED	PERIODICALLY DURING TASK LISTED
1. Observe drilling operations and maintain complete and accurate records for each pier.	X	—
2. Verify placement locations and plumbness, confirm pier diameters, bell diameters (if applicable), lengths, embedment into bedrock (if applicable) and adequate end bearing strata capacity.	X	—
3. For concrete piers, perform additional inspections in accordance with Section 1704.4.	—	—
4. For masonry piers, perform additional inspections in accordance with Section 1704.5.	—	—

ASTM E 605, the standard reference. The thickness is determined by taking the average of not less than four measurements for each 1,000 square feet of the sprayed area on each floor or part thereof. ASTM E 605 also determines the thickness of spray for structural framing members, and testing is performed on no less than 25 percent of the structural members on each floor.

Mastic and Intumescent Coatings

This section deals with mastic and intumescent fire-resistant coatings that are applied over structural elements, decks, and exterior insulation and finish systems (EIFS). Special inspections do not apply to exterior insulation and finish systems installed over a water-resistive barrier that has a means of draining moisture to the exterior or to exterior insulation and finish systems installed over masonry or concrete walls.

Any proposed work that is, in the opinion of the building official, unusual in its nature will be subject to special inspections. Such work includes the following:

• Construction materials and systems that are alternatives to materials and systems prescribed by the code

• Unusual design applications of materials described in this code

• Materials and systems required to be installed in accordance with additional manufacturer's instructions that set down requirements not contained in the code or in the standards reference by the code

!Definitionalert

Special Inspection, Periodic: The part-time or intermittent observation of work requiring special inspection by an approved special inspector who is present in the area where the work has been or being performed and at the completion of the work.

Statement of Special Inspections

A statement of special inspections must identify the following:

• The materials, systems, components, and work required to have special inspection or testing by the building official or by the registered design professional responsible for each portion of the work

• The type and extent of each special inspection

• The type and extent of each test

• Additional requirements for special inspection or testing for seismic or wind resistance as specified in this chapter

• For each type of special inspection, identification as to whether it will be continuous special inspection or periodic inspection

The first case regarding a statement of special inspection that I want to go over is for seismic resistance. The statement has to include seismic requirements for the following cases:

• The seismic-force-resisting systems in structures assigned to Seismic Design Category C, D, E, or F in accordance with Chapter 16

• Designated seismic systems in structures assigned to Seismic Design Category D, E, or F

!Definitionalert

Special Inspection, Continuous: The full-time observation of work requiring special inspection by an approved special inspector who is present in the area where the work is being performed.

!Codealert

The special inspections for steel elements of buildings and structures shall be as required by Section 1704.3 and Table 1704.3.

- The following additional systems and components in structures assigned to Seismic Design Category C and D: heating, ventilating, and air-conditioning (HVAC) ductwork containing hazardous materials and anchorage of such ductwork
- Piping systems and mechanical units containing flammable, combustible, or highly toxic materials
- Anchorage of electrical equipment used for emergency or standby power systems
- The following additional systems and components in structures assigned to Seismic Design Category D are found below:
  - Systems required for Seismic Design Category C
  - Exterior-wall panels and their anchorage
  - Suspended-ceiling systems and their anchorage
  - Access floors and their anchorage
  - Steel storage racks and their anchorage whose importance factor is equal to 1.5 or in accordance with ASCE 7
- The following additional systems and components in structures assigned to Seismic Design Category E or F:
  - Systems required for Seismic Design Categories C and D
  - Electrical equipment

There is a change in one of the exceptions from the previous code regarding detached one- or two-family dwellings that are not more than two stories in height and that do not have torsional, nonparallel, or stiffness irregularities or discontinuity in capacity.

The statements of special inspections for seismic conditions are strict and must include the designated seismic and seismic-force-resisting systems and any additional special inspections that are required by this chapter.

In Chapter 16, I discussed wind loads for buildings and structures. Buildings and structures are required to have a statement of special inspection for structures constructed in areas where the 3-second-gust basic wind speed is 120 mph, which is considered to be Exposure Category B. I Exposure Category C or D, the 3-second-gust basic wind speed is 110 mph. The statement of special inspection for wind speed must include roof cladding, wall connections to the roof, and the floor diaphragm and win-force-resisting-system connections to the foundation. This rule does not apply to fabrication of manufactured systems or components that have a label that indicates compliance with the wind-load and impact-resistance requirements.

Contractor Responsibility

If you are a contractor and are responsible for the construction of a main wind- or seismic-force-resisting system, designated seismic system, or a wind- or seismic-resisting component listed in the statement of special inspections, you must submit a statement claiming responsibility and give this to the building official and to the owner. This must be done before you begin any work on the system. When writing your statement, you must include the following information:

• An acknowledgment that you are aware of the special requirements that are contained in the statement of special inspections

• An acknowledgment that control will be exercised to obtain conformance with the construction documents approved by the building official

• Procedures that you will use to exercise control within the contractor's organization, the method and the frequency of reporting, and the distribution of the reports

• Identification and qualifications of the person(s) exercising this control and their position(s) in the organization

DESIGN STRENGTHS OF MATERIALS

The building official will confirm the strength and stress grade of any structural material and make the decision whether they conform to the specifications and methods of design in accepted engineering practice or the approved rules in the absence of applicable standards for new materials that are not specifically mentioned in this code.

!Codealert

Sections 1705 and 1706 contain a number of changes that should be observed for code compliance.

If there are no approved rules or other standards, it will be up to the building official to make any tests or investigations. The building official can accept authenticated reports from approved agencies as well. Let me remind you that the financial responsibility for any required testing belongs to the permit holder.

IN-SITU LOAD TESTS

There may be a time when the stability or load-bearing capacity of a building or structure is questioned. This can be very frustrating, given all the hard work, time, and effort it took to create your building, which may now not pass inspection. If this event should indeed take place, an engineering test is mandatory and includes a structural analysis, an in-situ load test, or possibly both.

The structural analysis is based on actual material properties and other as-built conditions that affect the stability or load-bearing capacity. Such testing must be done in accordance with the applicable design standard. In the event that the results are not as positive as you would like them do be, modifications must be made to ensure structural adequacy or to remove the inadequate construction.

Chapter 35 contains material standards that you can consult to determine which you must follow. In-situ load tests are to be conducted in accordance with this chapter and must be supervised by a registered design professional. The test must simulate the real test so to address the concerns regarding the structural stability of your building or structure.

TEST STANDARDS FOR JOIST HANGERS AND CONNECTORS

The last section concerns test standards for joist hangers and connectors. The vertical load-bearing capacity, torsional-moment capacity, and deflection characteristics of joist hangers must be determined in accordance with ASTM D 1761 using lumber that has a specific gravity of 0.49 or greater but no greater than 0.55.

!Codealert
Structural observations in Section 1709 have changed, so check this section to get up to speed with the current code requirements.

> # !Codealert
>
> See all of Section 1704 for numerous code amendments.

The joist length is not required to be longer than 24 inches. The allowable vertical load of the joist hanger must be the lowest value determined by the following:

- The lowest ultimate vertical load for a single hanger from any test divided by three (where three tests are conducted and each ultimate vertical load does not vary more than 20 percent from the average ultimate vertical load)
- The average ultimate vertical load for a single hanger from all tests divided by three (where six or more tests are conducted)
- The average from all tests of the vertical loads that produce a vertical movement of the joist with respect to the header of 0.125
- The sum of the allowable design loads for nails or other fasteners utilized to secure the joist hanger to the wood members and allowable bearing loads that contribute to the capacity of the hanger
- The allowable design load for the wood members forming the connection

> # !Codealert
>
> A main wind-force-resisting system is an assemblage of structural elements assigned to provide support and stability for the overall structure. The system generally receives wind loading from more than one surface.

CHAPTER 18
SOILS AND FOUNDATIONS

This chapter contains provisions relating to building and foundation systems in areas that are not subject to scour or water pressure by the actions of wind or waves. Throughout this chapter there are many references to Chapter 16, which contains the provisions for buildings and foundations affected by scour or water pressure loads.

In this chapter there are allowable bearing pressures, stresses, and design formulas that are used with the allowable-stress-design-load specifications that are found in Chapter 16. When looking at the quality and design of materials used structurally in excavations, footings, and foundations, you must also look at Chapters 19, 21, 22, and 23, which contain requirements that the design materials used for excavations, footings, and foundations must conform to. Excavations and fills must also comply with Chapter 33 on construction safeguards.

If the foundation is proportioned using the load combinations of Chapter 16 and the computation of the seismic overturning moment is by the equivalent lateral-force method, the proportioning must be in accordance with Section 12.13.4 of ASCE 7.

FOUNDATION AND SOIL INVESTIGATIONS

This section contains provisions for foundation and soil investigations. Classification and investigation of the soil must be made by a registered design professional where required by the building official. If such investigation is required, it is the owner's or applicant's responsibility to submit the foundation and soil investigation to the building official.

There are many reasons why a building official may require a foundation and soil investigation. Any soils about which the classification, strength, or compressibility is in doubt or where a load-bearing value greater than that specified in the code is claimed, the building official will require that the necessary investigation to be made.

Soils that swell when subjected to moisture are classified as expansive soils and contain clay materials that attract and absorb water. If the building official suspects that you may have this type of soil, you can be sure that a soil test will be required.

There are four provisions that a soil must meet to be considered expansive. Tests that show compliance with the first three items will not be required if the test prescribed in item 4 is conducted:

• Plasticity index (PI) of 15 or greater, determined in accordance with ASTM D 4318

• More than 10 percent of the soil particles pass through a No. 2 sieve, determined in accordance with ASTM D 422

• More than 10 percent of the soil particles are less than 5 micrometers in size, also determined in accordance with ASTM D 422

• Expansion index greater than 20, determined in accordance with ASTM D 4829

A subsurface soil investigation will be performed to find out if the groundwater table is above or within 5 feet of the elevation of the lowest floor level where floors are located below the finished ground level adjacent to the foundation. A subsurface soil investigation is not required if you can provide waterproofing that is in accordance with the provisions of this chapter.

Another reason for foundation and soils investigations is for buildings and structures with a Seismic Design Category of C, D, E, or F. The investigation for category C differs from D, E, and F and includes an evaluation of the potential hazards that result from earthquake motions such as slope instability, liquefaction, and surface rupture due to faulting or lateral spreading. Categories D, E, and F have the same investigation as category C in addition to the following:

• To determine the lateral pressures that are made by earthquake motions and the effect they have on basement and retaining walls

• An assessment of potential consequences of any liquefaction and loss of soil strength, including estimation of differential settlement, lateral movement, or reduction in foundation and mitigation measures; you may use these measures in the design of the structure and can include ground stabilization and selection of appropriate foundation types

• An evaluation for liquefaction and soil-strength loss for site peak ground acceleration; an exception is a site-specific study where peak ground acceleration is determined in accordance with Section 21.2.1 of ASCE 7

• Foundation and soils investigations including soil classification according to general behavior under given physical conditions

Table 18.1 shows the different classes of soil, the allowable foundations, and the lateral bearings of each class.

Necessary tests of materials from borings, test pits, or other subsurface exploration and observation are used to determine soil classification. If required, additional studies must be made to evaluate the following:

- Slope stability
- Soil strength
- Position and adequacy of load-bearing soils
- Effect of moisture variation
- Compressibility
- Liquefaction and expansiveness

TABLE 18.1 Footings supporting walls of light-frame construction.[a,b,c,d,e]

CLASS OF MATERIALS	ALLOWABLE FOUNDATION PRESSURE (psf)[d]	LATERAL BEARING (psf below natural grade)[d]	LATERAL SLIDING	
			Coefficient of friction[a]	Resistance (psf)[b]
1. Crystalline bedrock	12,000	1,200	0.70	—
2. Sedimentary and foliated rock	4,000	400	0.35	—
2. Sandy gravel and/or gravel (GW and GP)	3,000	200	0.35	—
3. Sand, silty sand, clayey sand, silty gravel and clayey gravel (SW, SP, SM, SC, GM and GC)	2,000	150	0.25	—
3. Clay, sandy clay, silty clay, clayey silt, silt and sandy silt (CL, ML, MH and CH)	1,500[c]	100	—	130

For SI: 1 pound per square foot = 0.0479 kPa, 1 pound per square foot per foot = 0.157 kPa/m.

a. Coefficient to be multiplied by the dead load.

b. Lateral sliding resistance value to be multiplied by the contact area, as limited by Section 1804.3.

c. Where the building official determines that in-place soils with an allowable bearing capacity of less than 1,500 psf are likely to be present at the site, the allowable bearing capacity shall be determined by a soils investigation.

d. An increase of one-third is permitted when using the alternate load combinations in Section 1605.3.2 that include wind or earthquake loads.

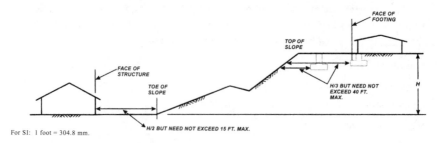

FIGURE 18.1 Foundation clearances from slopes.

The soil classification and design load-bearing capacity must be put on the construction document. The building official may request a written report of the investigation. The following list contains some of the information in the report:

- A plot showing the location of test borings and/or excavations
- A complete record of the soil samples
- A record of the soil profile
- Elevation of the water table, if encountered
- Recommendations for foundation type and design criteria
- Expected total and differential settlement
- Pole and pier foundation information
- Special design and construction provisions for footings or foundations founded on expansive soils, as necessary
- Compacted-fill material properties and testing

EXCAVATION, GRADING, AND FILL

Before digging any trenches, pits, tunnels, or other excavations, precautions must be taken so that lateral support from any footing or foundation is not disturbed or removed without unpinning or protecting the footing or foundation against settlement or lateral translation.

Backfilling is the refilling of an excavated space with soil free of organic material, construction debris, lumps that are larger than a pebble or filled with a controlled low-strength material (CLSM). Backfill must be compacted in a manner that does not damage the foundation or the water/dampproofing materials.

In areas where footings will bear on compacted fill areas, you must make sure that the compacted fill complies with the provisions of an approved report. Below is a list of the seven items that must be contained in the report:

- Specifications for the preparation of the site prior to placement of the compacted-fill material
- Specifications for material to be used as compacted fill
- Test method to be used to determine the maximum dry density and optimum moisture content of the material to be used as compacted fill
- Maximum allowable thickness of each lift of compacted-fill material
- Field test methods for determining the in-place dry density of the compacted fill
- Minimum acceptable in-place dry density expressed as a percentage of the maximum dry density determined in accordance with item 3 of this list
- Number and frequency of field tests required to determine compliance with item 6 of this list

Please note the following exception: compacted-fill material that is less than 12 inches in depth does not need to comply with an approved report, provided that it has been compacted to a minimum of 90 percent modified proctor. See ASTM D 1557 for more information regarding this exception. Also note that controlled low-strength materials do not need to be compacted.

The provisions for controlled low-strength material (CLSM) on which footings will bear must follow similar rules. There is an approved report in which specific provisions must be followed:

- Specifications for the preparation of the site prior to placement of CLSM
- Specifications for the CLSM
- Laboratory or field test method(s) to be used to determine the compressive strength or bearing capacity of the CLSM
- Test methods for determining the acceptance of the CLSM in the field
- Number and frequency of field tests required to determine compliance with item 4 of this list

!Definitionalert

Special Inspection: Inspection of the materials, installation, fabrication, erection, or placement of components and connections that need special expertise to ensure compliance with approved construction documents and referenced standards.

Immediately adjacent to the foundation is where site grading takes place. The slope cannot be any less than 1 unit vertical for every 20 units horizontal for a minimum distance of 10 feet measured perpendicular to the face of the wall. You must provide a 5 percent slope or an approved alternative if physical obstructions or lot lines prohibit the original 10-foot requirement. When using swales, they must be sloped for a minimum of 2 percent where located within 10 feet of the building foundation.

Resistant surfaces that are within 10 feet of the building foundation must be sloped for a minimum of 2 percent away from the building. One exception to this rule is for climatic or soil conditions. If such conditions exist, the slope of the ground away from the building foundation is allowed to be reduced, but no less than 1 unit vertical in 48 units horizontal (2 percent slope). The procedure that you use to establish the final ground level will account for additional settlement of the backfill. Grading and fill for flood-hazard areas will not be approved unless the following is true:

• Fill is placed, compacted, and sloped to minimum shifting, slumping, and erosion during the rise and fall of flood water and, if applicable, wave action.

• Demonstration through hydrologic and hydraulic analyses performed by a registered design professional has been done in accordance with standard engineering practice and shows that the proposed grading or fill will not result in any increase flood levels.

• Fill is conducted and/or placed to avoid diversion of water and waves toward any building or structure.

• Demonstration that the cumulative effect of the proposed flood-hazard area encroachment, when combined with all other existing and anticipated flood-hazard area encroachment, will not increase the design flood elevation more than 1 foot at any point.

Be sure you have any permits that you need for excavation or grading. If you're not sure, ask the building official and don't assume that the contractor has obtained them. You want to make sure that your construction and building plans go off without a hitch.

LOAD-BEARING VALUES OF SOILS

Table 18.1 provides the acceptable load-bearing values that are to be used with the allowable-stress design-load combinations as specified in Chapter 16. There are maximum allowances that you must adhere to for foundation pressure, lateral pressure, or lateral sliding-resistance values. These must not exceed the values found in Table 18.1 unless you have data to verify the use of a higher value. Any higher values must be submitted and approved for use.

The values found in Table 18.1 apply to any materials that have similar physical characteristics and nature. Do not assume that mud, organic silt, organic clays, peat, or unprepared fill have an acceptable load-bearing capacity unless you have the data to back that up. I believe we all know what happens when we assume something to be true. And it would be a great deal of time, money, and energy wasted if you wrongly assume that the use of a material is okay without the data to back it up. That being said, there is, however, an exception to this. An acceptable load-bearing capacity is permitted to be used if the building official considers the load-bearing capacity of mud, organic silt, or unprepared fill adequate for the support of lightweight and temporary structures.

To determine the resistance of structural walls to lateral sliding, calculate by combining the values from the lateral bearing and sliding resistance that are found in Table 18.1, unless you have data to prove why you are using higher values. Remember you have to submit the reasons or data and obtain approval.

In the case of sandy or silty clays or clayey silt, under no circumstance can the lateral sliding resistance be more than one-half of the dead load. It is possible for increases to be allowed for lateral sliding resistance. For each additional foot of depth to a maximum of 15 times the tabular value (Table 18.1).

FOOTINGS AND FOUNDATIONS

Footings and foundations are crucial to the success of a building. When properly designed, footings and foundations withstand the forces that a building is subjected to. Footings and foundations are built directly on undisturbed soil, compacted-fill material, or CLSM, with a minimum depth of footings below the undisturbed surface of 12 inches. While the top surface of footings has to be level, the bottom surface does not. The bottom surface of footings is allowed to have a slope not to exceed 1 unit vertical in 10 units horizontal.

There are times when it is necessary to change the elevation of the top surface; in this case footings must be stepped. Another important element regarding footings and foundations is frost protection. Foundation walls, piers, and other permanent supports of buildings and structures must be protected by either extending below the frost line, erecting on solid rock, or constructing in accordance with ASCE 32. If your building is free-standing and all of the following conditions are met, frost protection is not required.

An area of 600 square feet or less classified in Occupancy Category I as light-frame construction or of 400 square feet or less for other than light-frame constructions and with an eave height of 10 feet or less, as specified in Chapter 16, does not require frost protection.

Footings that are on granular soils (soils consisting mainly of sands and gravels) must be located so that the line drawn between the lower edge of adjoining footings will not have a steeper slope than 30 degrees with the horizontal, unless the material supporting the higher footing is braced or retained or otherwise laterally supported in a approved manner. For the most part, buildings that are below slopes are set apart from the slope at an acceptable distance from the slope for protection from slope drainage. Figure 18.1 contains satisfactory distances of foundation clearances from slopes.

In cases where the existing slope is steeper than one unit vertical in one unit horizontal, the toe of the slope is assumed to be at the foundation.

Most wet basements are caused by surface water that has not adequately drained from the foundation wall. On graded sites, the top of the exterior foundation is to extend above the elevation of the street gutter from an approved drainage device. Alternate elevations are allowed, with permission of the building official, if you can demonstrate that required drainage to the point of discharge and away from the structure is provided at all locations.

An alternate setback and clearance are also allowed, once again, with the building official's permission. The official has the authority to request an investigation and recommend a registered design professional to demonstrate that the intent has been satisfied. Any investigation includes consideration of the material, height of the slope and slope gradient, load intensity, and erosion characteristics of any materials used for the slope.

Footings are required to follow code regarding design, design load, vibratory loads, and much more. First, let's take a look at footing designs: footings must be designed so that the allowable bearing capacity of the soil is not exceeded and the differential settlement is minimized.

Footings are to have a minimum width of 12 inches. Chapter 16 contains provisions for unfavorable effects due to the combinations of loads, and footings must be designed with this in mind. The dead load is permitted to include the weight of foundations, footings, and overlying fill. Reduced live loads, as specified in Chapter 16, are allowed to be used in the design of footings.

When machinery-operation or other vibrations are sent through the foundation, you must give consideration in the footing design to prevent detrimental disturbances of the soil. From vibratory loads to concrete footing there is much to learn and understand. Design, materials, and construction of concrete footings not only have to comply with this code chapter but also Chapter 19, which is about concrete. Please be aware, though, that in instances where a specific design is not provided, concrete footings that support walls that are made of light-frame construction can be designed in accordance with Table 18.3, and all concrete in footings must have a specified compressive strength of no less than 2,500 pounds per square inch at 28 days.

Plain concrete footings that support walls for buildings other than light-frame construction cannot have an edge thickness of less than 8 inches placed on soil. The exception is that plain concrete footings that support Group R-3 occupancies can have an edge thickness of 6 inches but only when the footing does not extend beyond the thickness of the footing on either side.

It is important that you do not place concrete footings through water. You may seek permission from the building official if you are using a funnel or some other method. The key is to get the building official to give you a thumbs-up to use a funnel to place concrete. You must be sure to protect concrete from freezing during the time that you are placing it and for a total of no less than 5 days thereafter. Under no circumstance is water allowed to flow through concrete that has been laid in the ground.

Both concrete and foundation walls must be designed in accordance with Chapters 19 and 21 with regard to concrete and masonry. Foundation walls that are supported laterally at the top and bottom and are within the guidelines of Tables 18.3 to 18.7 may be designed and constructed in accordance with this section.

Table 18.2 covers plain foundation walls; 18.3, 18.4, and 18.5, masonry walls; and 18.6; concrete foundation walls. These tables will prove to be very helpful when you are ready to face this challenge. Masonry must be laid in running board, and the mortar has to be Type M or S. Both are used for below-grade load-bearing masonry work.

When you have a difference in height between the exterior-finish ground level and the lower or the top of the concrete footing that supports the foundation, you end up with an unbalanced backfill height. There are instances when an unbalanced backfill height is permitted, such as where an interior concrete slab on grade is provided and is in contact with the interior surface of the foundation wall. Foundation walls of rough or random rubble stone cannot be less than 16 inches thick, and rubble stone cannot be used for foundations for structures in Seismic Design Category C, D, E, or F.

!Codealert

Foundation design for seismic overturning refers to the requirement that the foundation be proportioned using the load combinations of Section 1605.2. The computation of the seismic overturning moment is by the equivalent lateral-force method or the modal analysis method, and the proportioning shall be in accordance with Section 12.13.4 or ASCE 7.

TABLE 18.2 Plain masonry foundation walls.[a,b,c]

MAXIMUM WALL HEIGHT (feet)	MAXIMUM UNBALANCED BACKFILL HEIGHT[e] (feet)	MINIMUM NOMINAL WALL THICKNESS (inches)		
		Soil classes and lateral soil load[a] (psf per foot below natural grade)		
		GW, GP, SW and SP soils 30	GM, GC, SM, SM-SC and ML soils 45	SC, ML-CL and Inorganic CL soils 60
7	4 (or less)	8	8	8
	5	8	10	10
	6	10	12	10 (solid[c])
	7	12	10 (solid[c])	10 (solid[c])
8	4 (or less)	8	8	8
	5	8	10	12
	6	10	12	12 (solid[c])
	7	12	12 (solid[c])	Note d
	8	10 (solid[c])	12 (solid[c])	Note d
9	4 (or less)	8	8	8
	5	8	10	12
	6	12	12	12 (solid[c])
	7	12 (solid[c])	12 (solid[c])	Note d
	8	12 (solid[c])	Note d	Note d
	9	Note d	Note d	Note d

For SI: 1 inch = 25.4 mm, 1 foot = 304.8 mm, 1 pound per square foot per foot = 0.157kPa/m.

a. For design lateral soil loads, see Section 1610. Soil classes are in accordance with the Unified Soil Classification System and design lateral soil loads are for moist soil conditions without hydrostatic pressure.

b. Provisions for this table are based on construction requirements specified in Section 1805.5.2.2.

c. Solid grouted hollow units or solid masonry units.

d. A design in compliance with Chapter 21 or reinforcement in accordance with Table 1805.5(2) is required.

e. For height of unbalanced backfill, see Section 1805.5.1.2.

Foundation-wall materials such as concrete must be constructed in accordance with Table 18.6 and with the following list:

• The size and spacing of vertical reinforcement shown in Table 18.6 is based on the use of reinforcement with minimum yield strength of 60,000 psi or vertical reinforcement with minimum yield strength of 40,000 psi; 50,000 psi is permitted, provided the same size bar is used and the spacing shown in the table is reduced by multiplying the spacing by 0.67 or 0.83.

• Vertical reinforcement, when required, must be placed nearest the inside face of the wall a distance from the outside face of the wall. The distance is equal to the

TABLE 18.3 Eight-inch masonry foundation walls with reinforcement where d ≥ 5 inches.[a,b,c]

MAXIMUM WALL HEIGHT (feet-inches)	MAXIMUM UNBALANCED BACKFILL HEIGHT[d] (feet-inches)	VERTICAL REINFORCEMENT		
		Soil classes and lateral soil load[a] (psf per foot below natural grade)		
		GW, GP, SW and SP soils 30	GM, GC, SM, SM-SC and ML soils 45	SC, ML-CL and Inorganic CL soils 60
7-4	4-0 (or less)	#4 at 48" o.c.	#4 at 48" o.c.	#4 at 48" o.c.
	5-0	#4 at 48" o.c.	#4 at 48" o.c.	#4 at 48" o.c.
	6-0	#4 at 48" o.c.	#5 at 48" o.c.	#5 at 48" o.c.
	7-4	#5 at 48" o.c.	#6 at 48" o.c.	#7 at 48" o.c.
8-0	4-0 (or less)	#4 at 48" o.c.	#4 at 48" o.c.	#4 at 48" o.c.
	5-0	#4 at 48" o.c.	#4 at 48" o.c.	#4 at 48" o.c.
	6-0	#4 at 48" o.c.	#5 at 48" o.c.	#5 at 48" o.c.
	7-0	#5 at 48" o.c.	#6 at 48" o.c.	#7 at 48" o.c.
	8-0	#5 at 48" o.c.	#6 at 48" o.c.	#7 at 48" o.c.
8-8	4-0 (or less)	#4 at 48" o.c.	#4 at 48" o.c.	#4 at 48" o.c.
	5-0	#4 at 48" o.c.	#4 at 48" o.c.	#5 at 48" o.c.
	6-0	#4 at 48" o.c.	#5 at 48" o.c.	#6 at 48" o.c.
	7-0	#5 at 48" o.c.	#6 at 48" o.c.	#7 at 48" o.c.
	8-8	#6 at 48" o.c.	#7 at 48" o.c.	#8 at 48" o.c.
9-4	4-0 (or less)	#4 at 48" o.c.	#4 at 48" o.c.	#4 at 48" o.c.
	5-0	#4 at 48" o.c.	#4 at 48" o.c.	#5 at 48" o.c.
	6-0	#4 at 48" o.c.	#5 at 48" o.c.	#6 at 48" o.c.
	7-0	#5 at 48" o.c.	#6 at 48" o.c.	#7 at 48" o.c.
	8-0	#6 at 48" o.c.	#7 at 48" o.c.	#8 at 48" o.c.
	9-4	#7 at 48" o.c.	#8 at 48" o.c.	#9 at 48" o.c.
10-0	4-0 (or less)	#4 at 48" o.c.	#4 at 48" o.c.	#4 at 48" o.c.
	5-0	#4 at 48" o.c.	#4 at 48" o.c.	#5 at 48" o.c.
	6-0	#4 at 48" o.c.	#5 at 48" o.c.	#6 at 48" o.c.
	7-0	#5 at 48" o.c.	#6 at 48" o.c.	#7 at 48" o.c.
	8-0	#6 at 48" o.c.	#7 at 48" o.c.	#8 at 48" o.c.
	9-0	#7 at 48" o.c.	#8 at 48" o.c.	#9 at 48" o.c.
	10-0	#7 at 48" o.c.	#9 at 48" o.c.	#9 at 48" o.c.

For SI: 1 inch = 25.4 mm, 1 foot = 304.8 mm, 1 pound per square foot per foot = 0.157kPa/m.

a. For design lateral soil loads, see Section 1610. Soil classes are in accordance with the Unified Soil Classification System and design lateral soil loads are for moist soil conditions without hydrostatic pressure.

b. Provisions for this table are based on construction requirements specified in Section 1805.5.2.2.

c. For alternative reinforcement, see Section 1805.5.3.

d. For height of unbalanced backfill, see Section 1805.5.1.2.

TABLE 18.4 Ten-inch masonry foundation walls with reinforcement where d ≥ 6.75 inches.[a,b,c]

MAXIMUM WALL HEIGHT (feet-inches)	MAXIMUM UNBALANCED BACKFILL HEIGHT[d] (feet-inches)	VERTICAL REINFORCEMENT		
		Soil classes and lateral soil load[a] (psf per foot below natural grade)		
		GW, GP, SW and SP soils 30	GM, GC, SM, SM-SC and ML soils 45	SC, ML-CL and Inorganic CL soils 60
7-4	4-0 (or less)	#4 at 56" o.c.	#4 at 56" o.c.	#4 at 56" o.c.
	5-0	#4 at 56" o.c.	#4 at 56" o.c.	#4 at 56" o.c.
	6-0	#4 at 56" o.c.	#4 at 56" o.c.	#5 at 56" o.c.
	7-4	#4 at 56" o.c.	#5 at 56" o.c.	#6 at 56" o.c.
8-0	4-0 (or less)	#4 at 56" o.c.	#4 at 56" o.c.	#4 at 56" o.c.
	5-0	#4 at 56" o.c.	#4 at 56" o.c.	#4 at 56" o.c.
	6-0	#4 at 56" o.c.	#4 at 56" o.c.	#5 at 56" o.c.
	7-0	#4 at 56" o.c.	#5 at 56" o.c.	#6 at 56" o.c.
	8-0	#5 at 56" o.c.	#6 at 56" o.c.	#7 at 56" o.c.
8-8	4-0 (or less)	#4 at 56" o.c.	#4 at 56" o.c.	#4 at 56" o.c.
	5-0	#4 at 56" o.c.	#4 at 56" o.c.	#4 at 56" o.c.
	6-0	#4 at 56" o.c.	#4 at 56" o.c.	#5 at 56" o.c.
	7-0	#4 at 56" o.c.	#5 at 56" o.c.	#6 at 56" o.c.
	8-8	#5 at 56" o.c.	#7 at 56" o.c.	#8 at 56" o.c.
9-4	4-0 (or less)	#4 at 56" o.c.	#4 at 56" o.c.	#4 at 56" o.c.
	5-0	#4 at 56" o.c.	#4 at 56" o.c.	#4 at 56" o.c.
	6-0	#4 at 56" o.c.	#5 at 56" o.c.	#5 at 56" o.c.
	7-0	#4 at 56" o.c.	#5 at 56" o.c.	#6 at 56" o.c.
	8-0	#5 at 56" o.c.	#6 at 56" o.c.	#7 at 56" o.c.
	9-4	#6 at 56" o.c.	#7 at 56" o.c.	#8 at 56" o.c.
10-0	4-0 (or less)	#4 at 56" o.c.	#4 at 56" o.c.	#4 at 56" o.c.
	5-0	#4 at 56" o.c.	#4 at 56" o.c.	#4 at 56" o.c.
	6-0	#4 at 56" o.c.	#5 at 56" o.c.	#5 at 56" o.c.
	7-0	#5 at 56" o.c.	#6 at 56" o.c.	#7 at 56" o.c.
	8-0	#5 at 56" o.c.	#7 at 56" o.c.	#8 at 56" o.c.
	9-0	#6 at 56" o.c.	#7 at 56" o.c.	#9 at 56" o.c.
	10-0	#7 at 56" o.c.	#8 at 56" o.c.	#9 at 56" o.c.

For SI: 1 inch = 25.4 mm, 1 foot = 304.8 mm, 1 pound per square foot per foot = 0.157kPa/m.

a. For design lateral soil loads, see Section 1610. Soil classes are in accordance with the Unified Soil Classification System and design lateral soil loads are for moist soil conditions without hydrostatic pressure.

b. Provisions for this table are based on construction requirements specified in Section 1805.5.2.2.

c. For alternative reinforcement, see Section 1805.5.3.

d. For height of unbalanced backfill, see Section 1805.5.1.2.

TABLE 18.5 Twelve-inch masonry foundation walls with reinforcement where d ≥ 8.75 inches.[a,b,c]

MAXIMUM WALL HEIGHT (feet-inches)	MAXIMUM UNBALANCED BACKFILL HEIGHT[d] (feet-inches)	VERTICAL REINFORCEMENT		
		Soil classes and lateral soil load[a] (psf per foot below natural grade)		
		GW, GP, SW and SP soils 30	GM, GC, SM, SM-SC and ML soils 45	SC, ML-CL and Inorganic CL soils 60
7-4	4-0 (or less)	#4 at 72" o.c.	#4 at 72" o.c.	#4 at 72" o.c.
	5-0	#4 at 72" o.c.	#4 at 72" o.c.	#4 at 72" o.c.
	6-0	#4 at 72" o.c.	#4 at 72" o.c.	#5 at 72" o.c.
	7-4	#4 at 72" o.c.	#5 at 72" o.c.	#6 at 72" o.c.
8-0	4-0 (or less)	#4 at 72" o.c.	#4 at 72" o.c.	#4 at 72" o.c.
	5-0	#4 at 72" o.c.	#4 at 72" o.c.	#4 at 72" o.c.
	6-0	#4 at 72" o.c.	#4 at 72" o.c.	#5 at 72" o.c.
	7-0	#4 at 72" o.c.	#5 at 72" o.c.	#6 at 72" o.c.
	8-0	#5 at 72" o.c.	#6 at 72" o.c.	#7 at 72" o.c.
8-8	4-0 (or less)	#4 at 72" o.c.	#4 at 72" o.c.	#4 at 72" o.c.
	5-0	#4 at 72" o.c.	#4 at 72" o.c.	#4 at 72" o.c.
	6-0	#4 at 72" o.c.	#4 at 72" o.c.	#5 at 72" o.c.
	7-0	#4 at 72" o.c.	#5 at 72" o.c.	#6 at 72" o.c.
	8-8	#5 at 72" o.c.	#7 at 72" o.c.	#8 at 72" o.c.
9-4	4-0 (or less)	#4 at 72" o.c.	#4 at 72" o.c.	#4 at 72" o.c.
	5-0	#4 at 72" o.c.	#4 at 72" o.c.	#4 at 72" o.c.
	6-0	#4 at 72" o.c.	#5 at 72" o.c.	#5 at 72" o.c.
	7-0	#4 at 72" o.c.	#5 at 72" o.c.	#6 at 72" o.c.
	8-0	#5 at 72" o.c.	#6 at 72" o.c.	#7 at 72" o.c.
	9-4	#6 at 72" o.c.	#7 at 72" o.c.	#8 at 72" o.c.
10-0	4-0 (or less)	#4 at 72" o.c.	#4 at 72" o.c.	#4 at 72" o.c.
	5-0	#4 at 72" o.c.	#4 at 72" o.c.	#4 at 72" o.c.
	6-0	#4 at 72" o.c.	#5 at 72" o.c.	#5 at 72" o.c.
	7-0	#4 at 72" o.c.	#6 at 72" o.c.	#6 at 72" o.c.
	8-0	#5 at 72" o.c.	#6 at 72" o.c.	#7 at 72" o.c.
	9-0	#6 at 72" o.c.	#7 at 72" o.c.	#8 at 72" o.c.
	10-0	#7 at 72" o.c.	#8 at 72" o.c.	#9 at 72" o.c.

For SI: 1 inch = 25.4 mm, 1 foot = 304.8 mm, 1 pound per square foot per foot = 0.157kPa/m.

a. For design lateral soil loads, see Section 1610. Soil classes are in accordance with the Unified Soil Classification System and design lateral soil loads are for moist soil conditions without hydrostatic pressure.

b. Provisions for this table are based on construction requirements specified in Section 1805.5.2.2.

c. For alternative reinforcement, see Section 1805.5.3.

d. For height of unbalanced backfill, see Section 1805.5.1.2.

TABLE 18.6 Concrete foundation walls.[a,b,c]

MAXIMUM WALL HEIGHT (feet)	MAXIMUM UNBALANCED BACKFILL HEIGHT[d] (feet)	VERTICAL REINFORCEMENT								
		Design lateral soil load[a] (psf per foot of depth)								
		30			45			60		
		Minimum wall thickness (inches)								
		7.5	9.5	11.5	7.5	9.5	11.5	7.5	9.5	11.5
5	4	PC	PC	PC	PC	PC	PC	PC	PC	PC
	5	PC	PC	PC	PC	PC	PC	PC	PC	PC
6	4	PC	PC	PC	PC	PC	PC	PC	PC	PC
	5	PC	PC	PC	PC	PC	PC	PC	PC	PC
	6	PC	PC	PC	PC	PC	PC	PC	PC	PC
7	4	PC	PC	PC	PC	PC	PC	PC	PC	PC
	5	PC	PC	PC	PC	PC	PC	PC	PC	PC
	6	PC	PC	PC	PC	PC	PC	#5 at 48"	PC	PC
	7	PC	PC	PC	#5 at 46"	PC	PC	#6 at 47"	PC	PC
8	4	PC	PC	PC	PC	PC	PC	PC	PC	PC
	5	PC	PC	PC	PC	PC	PC	PC	PC	PC
	6	PC	PC	PC	PC	PC	PC	#5 at 43"	PC	PC
	7	PC	PC	PC	#5 at 41"	PC	PC	#6 at 43"	PC	PC
	8	#5 at 47"	PC	PC	#6 at 43"	PC	PC	#6 at 32"	#6 at 44"	PC
9	4	PC	PC	PC	PC	PC	PC	PC	PC	PC
	5	PC	PC	PC	PC	PC	PC	PC	PC	PC
	6	PC	PC	PC	PC	PC	PC	#5 at 39"	PC	PC
	7	PC	PC	PC	#5 at 37"	PC	PC	#6 at 38"	#5 at 37"	PC
	8	#5 at 41"	PC	PC	#6 at 38"	#5 at 37"	PC	#7 at 39"	#6 at 39"	#4 at 48"
	9[d]	#6 at 46"	PC	PC	#7 at 41"	#6 at 41"	PC	#7 at 31"	#7 at 41"	#6 at 39"
10	4	PC	PC	PC	PC	PC	PC	PC	PC	PC
	5	PC	PC	PC	PC	PC	PC	PC	PC	PC
	6	PC	PC	PC	PC	PC	PC	#5 at 37"	PC	PC
	7	PC	PC	PC	#6 at 48"	PC	PC	#6 at 35"	#6 at 48"	PC
	8	#5 at 38"	PC	PC	#7 at 47"	#6 at 47"	PC	#7 at 35"	#7 at 48"	#6 at 45"
	9[d]	#6 at 41"	#4 at 48"	PC	#7 at 37"	#7 at 48"	#4 at 48"	#6 at 22"	#7 at 37"	#7 at 47"
	10[d]	#7 at 45"	#6 at 45"	PC	#7 at 31"	#7 at 40"	#6 at 38"	#6 at 22"	#7 at 30"	#7 at 38"

For SI: 1 inch = 25.4 mm, 1 foot = 304.8 mm, 1 pound per square foot per foot = 0.157kPa/m.

a. For design lateral soil loads, see Section 1610. Soil classes are in accordance with the Unified Soil Classification System and design lateral soil loads are for moist soil conditions without hydrostatic pressure.

b. Provisions for this table are based on construction requirements specified in Section 1805.5.2.2.

c. For alternative reinforcement, see Section 1805.5.3.

d. For height of unbalanced backfill, see Section 1805.5.1.2.

!Definitionalert

Flexural Length: The length of the pile from the first point of zero lateral deflection to the underside of the pile cap or grade beam.

Micropiles: 12-inch-diameter or less bored, grouted-in-place piles incorporating steel pipe (casing) and/or steel reinforcement.

Pier Foundations: Isolated masonry or cast-in-place concrete structural elements extending into firm materials.

Piers: Elements relatively short in comparison to their width, with lengths less than or equal to 12 times the least horizontal dimension. Piers derive their load-carrying capacity through skin friction, end bearing, or a combination of both.

Pile Foundations: Concrete, wood, or steel structural elements either driven into the ground or cast in place.

Piles: Elements relatively slender in comparison to their length, with lengths exceeding 12 times the least horizontal dimension. [run in]Piles derive their load-carrying capacity through skin friction, end bearing, or a combination of both.

wall thickness minus 1 1/4 inches plus one-half the bar diameter. The reinforcement must be placed within a tolerance of 3/8 inch where the distance is less than or equal to 8 inches.

• Instead of the support shown in Table 8.6, smaller reinforcing-bar sizes with closer spaces can be used if this provides an equal amount of reinforcement.

• Any concrete covering used for reinforcement that is measured from the inside face of the wall cannot be less than 1/2 inch. If measured from the inside face of the wall, the measurement cannot be less than 1 1/2 inches for No. 5 bars and 2 inches for smaller bars.

• Concrete must have a specified compressive strength of not less than 2,500 psi at 28 days.

Just like concrete foundation walls, masonry foundation walls must comply with similar standards. Please refer to the following list:

• The minimum vertical reinforcement for masonry foundations has a strength of 60,000 psi.

• The specified location of the reinforcement has to be equal or greater to the depth distance in Tables 18.3, 18.4, and 18.5.

• Masonry units must be installed with Type M or S mortar.

!Codealert

Any substantial sudden increase in rate of penetration of a timber pile shall be investigated for possible damage. If the sudden increase in rate of penetration cannot be correlated to soil strata, the pile shall be removed for inspection or rejected.

DAMPPROOFING AND WATERPROOFING

Dampproofing and waterproofing are both designed to resist moisture. Waterproofing is defined as the treatment of a structure or surface to protect the passage of water under pressure, while dampproofing is the treatment of a structure or surface to resist the passage of water (not pressurized).

Any walls, interior spaces, and floors below grade level must be water- and dampproofed. A foundation drain must be installed around the portion of the perimeter where the basement floor is below ground level. There are situations in which you only need to dampproof floors and walls, such as when the ground-water table is lowered and maintained at an elevation not less than 6 inches below the bottom of the lowest floor. You need to design your system to lower the ground-water table and base it on accepted principles of engineering that consider the permeability of the soil or rate of water entering the system.

In times where hydrostatic pressure does not occur, floors and walls for other than wood foundation systems are to be dampproofed. Check with AF&PA Technical Report No. 7 for regulations regarding construction of wood foundations.

You may think that it would be simpler to just dampproof by painting the interior walls with any type of material, but remember this: dampproofing will only retard moisture; it cannot stop water that is bearing against a foundation. Always install dampproofing materials between the floor and the base, except when a separated floor is installed above a concrete slab.

Dampproofing installed beneath the slab cannot consist of less than 6-mil polyethylene with joints lapped no less than 6 inches. If the code permits you to apply materials on top of the slab, they must consist of mopped-on bitumen and not less than 4-mil polyethylene. All joints in the membrane have to be lapped and sealed. Please check with the manufacturer's installation instructions.

Dampproofing walls is obviously going to be a bit different that dampproofing floors. When you apply dampproofing materials, make sure that you completely cover the exterior surface of the wall and extend it from the top of the footing to above the ground level. Your dampproofing must consist of a bituminous material of 3 pounds per square yard of acrylic-modified cement. You may use a surface-bonding mortar, but check with ASTM C 887 to assure that you are in compliance with the code. Be sure to seal any holes that you find in the concrete walls before you apply any dampproofing materials.

Any time a ground-water investigation reveals that a hydrostatic-pressure condition exists and the design does not include a ground-water-control system, all floors and walls must be waterproofed in accordance with this section. A wet basement is a common problem, but don't let it be your common problem.

Floors that are required to be waterproofed must be of concrete. Concrete floors must be designed and constructed to withstand the hydrostatic pressures that they will be faced with. By placing a membrane of rubberized asphalt, butyl rubber, or polyolefin composite you will accomplish this. As with dampproofing, you must make sure that all joints in the membrane are lapped and sealed in accordance with the manufacturer's installation instructions.

Walls are required to have waterproofing applied for the same reasons and require it to be applied from the bottom of the wall to no less than 12 inches above the maximum elevation of the ground-water table. You only need dampproof the rest of the wall in accordance to this chapter. You must waterproof with a two-ply hot-mopped felt, no less than 6-mil polyvinyl chloride, 40-mil polymer-modified asphalt, or another method that has been approved and is capable of bridging nonstructural cracks. Be sure that all joints in the membrane are lapped and sealed. Once again, check with the manufacturer's installation instructions. Also be sure to seal any joints in the walls and floors and penetrations of the floor and walls with a water-tight material. Always check with the code or the building official for approval of any methods and materials used.

If you are building in an area where hydrostatic pressure is nonexistent, you must provide dampproofing and install a base under the floor and a drain around the perimeter of the basement. If you have a subsoil drainage system that is designed and constructed in accordance with this chapter, it will be considered adequate for lowering the ground-water table. Let's take a look at the foundation drain requirements.

!Codealert
See Section 1810.8 for new code regulations on micropiles.

A drain has to have gravel or crushed stone, but no more than 10 percent of it must have the ability to pass through a No. 4 sieve or strainer. You must leave a minimum of 12 inches extending beyond the outside edge of the footing. Be sure that the bottom of the drain isn't higher than the bottom of the base under the floor and that the top of the drain isn't less than 6 inches above the top of the footing. Both the top of the drain and the top of the joints or of the perforations have to be covered with an approved filer membrane material. The floor base and foundation drain must discharge (by gravity or mechanical means) into an approved drainage system complying with the International Plumbing Code. The exception is that on a site that is located in a well-drained gravel or sand/gravel-mixture soil, a dedicated drainage system is not required. If you have questions regarding this exception, please contact your local code official.

PIER AND PILE FOUNDATIONS

I am starting this section with some definitions of common terms in this section:

The general requirements for piers and piles must follow the provisions of this section where Group R-3 and U occupancies do not exceed two stories of light-frame construction or where the surrounding foundation materials furnish adequate lateral support for the pile. These materials are subject to the approval of the building official. You must design and install your pier and pile foundations on the basis of a foundation investigation (as defined earlier in the chapter), unless you already have sufficient data. An investigation and report for pier and pile foundations include the following:

• Recommended pier or pile types and installed capacities

• Recommended center-to-center spacing of piers or piles

• Driving criteria

• Installation procedures

• Field inspection and reporting procedures (to include procedures for verification of the installed bearing capacity where required)

• Pier or pile load-test requirements

• Durability of pier or pile materials

• Designation of bearing stratum or strata

• Reductions for group action, where necessary

There are special types of piles that are not specifically mentioned in the code that can be used. However, you must submit acceptable test data, calculations, and other information that relates to the structural properties and load capacity of such piles. It is only after the building official reviews and approves such information that you will be allowed to use the special pile. In any case, the allowable stresses cannot exceed the limitations that have been set.

You must brace all piers and piles to provide lateral stability in all directions. To be considered braced, three or more piles must be connected by a rigid cap, provided that the piles are located in radial directions from the center of mass of the group and not less than 60 degrees apart. A two-pile group in a rigid cap is also considered to be braced along the axis connecting the two piles. Any methods that are used to brace piers or piles must be subject to the approval of the building official. Be careful when installing piers and piles.

Any disturbance in the required sequence of the installation can cause distortion and damage and adversely affect the structural integrity of piles that you are currently installing or that are already in place. The 2006 International Building Code will allow you to reuse existing piers or piles under certain circumstances and with the approval of the building official. You will have to submit evidence that the piers or piles are sound and meet all requirements of the code.

All existing piers and piles have to be load-tested or redriven to verify their capabilities. The design load applied to such piers or piles must be the lowest allowable load as determined by such tests. There is an approved formula to determine the allowable axial and lateral loads on piers or piles. The allowable compressive load on any pile where determined by the application of an approved driving formula cannot be more than 40 tons. You must use the wave-equation method of analysis for allowable loads over 40 tons to estimate both driving stresses and net displacement per blow at the ultimate load. To use a follower, you must obtain permission from the building official and you cannot use a fresh hammer cushion or pile-cushion material just prior to final penetration.

When there is any doubt regarding design load for any pier or pile foundation, testing must be done in accordance with ASTM D 1143 or ASTM D 4945. The following are allowable methods of load-test evaluations:

• Davisson offset limit

• Brinch-Hansen 90-percent criterion

• Butler-Hoy criterion

• Other methods approved by the building official

Piers, individual piles, and group of piles must develop ultimate load capacities of at least twice the design working loads in the designated load-bearing layers. And load-bearing capacities of piers or piles that are discovered to have a sharp or sweeping bend will be determined by an approved method of analysis or by load-testing a representative pier or pile. The maximum compressive load on any pier or pile due to mislocation cannot be more than 110 percent of the allowable design load. All piers and piles need proper lateral support to prevent buckling. This support must be in accordance with accepted engineering practice and provisions of the code. Unbraced piles in air, water, or fluid soils have to be designed as columns in accordance with the provisions of the code. Piles that are driven into firm ground are considered to be fixed and laterally supported at 5 feet below the ground surface; in soft material, at 10 feet. The building official does have the authority to make any alterations after a foundation investigation by an approved agency.

In addition to allowable pier and pile loads you must be aware of the requirements for seismic design. In the first part of this section you will learn about Seismic Design Category C. (Categories D, E, and F will follow shortly thereafter.) Individual pile caps, piers, or pile groups must be interconnected by ties. These ties must be capable of carrying a force equal to the product of the larger pile cap or column load times the seismic coefficient divided by 10. This can be disregarded only if you can demonstrate that equal restraint is provided with reinforced concrete beams within slabs on grade, reinforced concrete slabs on grade, or very dense granular soils. The code provides an exception: piers, supporting foundation walls, isolated interior posts detailed so the piers are not subject to lateral loads, lightly loaded exterior decks, and patios of Group R-3 and U occupancies not exceeding two stories of light-frame construction are not subject to interconnection if it can be shown that the soils are of adequate stiffness, subject to the approval of the building official.

Seismic-category structures must connect concrete piles and concrete-filled steel-pipe piles to the pile cap by embedding the pile reinforcement or by field-placed dowels anchored in the concrete pile. They must be embedded for a distance that is equal to the development length. For deformed bars the development length is the full development length for a compression or tension. You must be sure that the ends of hoops, spirals, and ties are terminated with seismic hooks.

The American Concrete Institute (ACI) 318 Building Code Requirements for Structural Concrete, Section 21.1 defines this with more clarity, and I recommend that you refer to it when terminating these ends. Please note that anchorage of concrete-steel pipe piles is allowed to be done using deformed bars developed into the concrete portion of the pile. Structures that are assigned to Seismic Design Category C must follow design details.

Pier or pile moments, shears, and lateral deflections used for design have to be established considering the nonlinear interaction of the shaft and soil, as recommended by a registered design professional. A pile may be assumed to be rigid if the ratio of the depth of embedment of the pile-to-pile diameter or width is less than or equal to six. You must always include pile-group effects from soil on lateral pile nominal strengths where pile center-to-center spacing in the direction of lateral force is less than eight pile diameters. The same is true for vertical pile strength where center-to-center spacing is less than three pile diameters. Did you know that provisions must be made so that specified lengths where a minimum length for reinforcement is specified must be maintained at the top of the pier or pile? It's true: provisions must be made so that those lengths or extents are maintained after pier or pile cutoff.

Seismic Design Categories D, E, and F must adhere to the requirements for Seismic Design Category C in addition to their own. Provisions of the American Concrete Institute (ACI) 318, Building Code Requirements for Structural

Concrete, Section 21.10.4 must apply when not in conflict with this chapter. Concrete for category D, E, or F must have a specified compressive strength of not less than 3,000 psi at 28 days. Please see the following list for exceptions to the above:

• Group R or U occupancies of light-framed construction and two stories or less in height are allowed to use concrete with a specified compressive strength of not less than 2,500 psi at 28 days.

• Detached one- and two-family dwellings of light-frame construction and two stories or less in height are not required to comply with the provisions of ACI 318, Section 21.10.4.

• Section 21.10.4 of ACI 318 does not apply to concrete piles.

The design details of piers, piles, and grade beams have to be designed and constructed to withstand maximum imposed curvatures from earthquake and ground motions and structure response. Curvatures must include free-field soil strains that have been modified for soil-pile-structure interaction. Site Class E or F sites have to be designed and detailed in accordance with ACI 318, Sections 21.4.4.1, 21.4.4.2, and 21.4.4.3 within seven pile diameters of the pile cap and the interfaces of soft to prestressed concrete piles. ACI 318 dictates many provisions regarding seismic design, including grade beams. However, grade beams that have the capacity to resist the forces from load combinations do not need to conform to ACI 318, Chapter 21. For more information regarding load combinations please refer to Chapter 16 of the 2006 International Building Code and to ACI 318 for any clarification regarding grade beams.

For piles that are required to resist uplift forces or provide rotational restraint, design of the anchorage of piles into the pile cap has to consider the combined effect of axial forces. The minimum of 25 percent of the strength of the pile in tension must include anchorage. Anchorage into the pile cap must be capable of developing the following:

• In the case of uplift, the lesser of the nominal tensile strength of the longitudinal reinforcement in a concrete pile, the nominal tensile strength of a steel pile, the pile uplift soil nominal strength factored by 1.3, or the axial tension force resulting from the load combinations in Chapter 16.

!Definitionalert

Aisle: An exit-access component that defines and provides a path of egress travel.

• In the case of rotational restraint, the lesser of the axial and shear forces and moments resulting from the load combinations of Chapter 16 or development of full axial and shear nominal strength of the pile.

If the vertical lateral-force-resisting elements are columns, the grade-beam or pile-cap flexural strengths must exceed the column flexural strength. The connections between batter piles and grade beams or pile caps must be designed to resist the nominal strength of the pile acting as a short column. Batter piles and connections must be capable of resisting forces and moments from the load combinations of Chapter 16.

Driven-Pile Foundations

Timber is strong, light in weight, and capable of adequate support. Timber piles are round, tapered timbers with the small end embedded into the soil. Timber piles used to support permanent structures are to be treated in accordance with this section and must be designed in accordance with AFPA NDS (American Forest and Paper Association). Round timber piles must conform to ASTM D 25, while sawn timber piles must conform to DOC PS-20 (Department of Congress). Timber piles that are used for support in permanent structures must comply with this section. If it is established that you will be using the tops of the untreated timber piles below the lowest ground-water level assumed to exist during the life of the lowest structure, timber piles do not have to comply with this section.

The AWPA U1 (Commodity Specifications E, Use Category 4C) contains very important information that you need to refer to for driven-pile foundations. If you are working with timber pile and suddenly notice an increase in rate of penetration, you must conduct an investigation for possible damage. If the sudden increase in rate of penetration is not related to soil strata, you must remove the pile for inspection and if nonviable the timber pile must be rejected.

The second type of driven-pile foundation is precast-concrete pile. Precast-concrete piles have to comply with design and manufacture requirements, must be of a minimum dimension, and must follow reinforcement and installation rules.

To resist all stresses brought on by handling, driving, and service loads, piles must be designed and manufactured in accordance with accepted engineering practices. Concrete piles must have a minimum lateral dimension of 8 inches; corners of square piles must be chamfered, which means to have a bevel or groove; and the longitudinal reinforcement must be at least 0.8 percent of the concrete section and consist of at least four bars. You must never drive a precast-concrete pile before the concrete has attained a compressive strength of at least 75 percent of the 28-day specified compressive strength, but no less than the strength sufficient to withstand handling and driving force.

Micropiles

Micropiles are 12-inch-diameter or less bored, grouted-in-place piles incorporating steel pipe (casing) and/or steel reinforcement. There has been a change in the code regarding micropiles. Keep your eyes open for any details that pertain to your construction or building needs and as always ask your local building official to clarify any questions that you may have.

Micropiles must have a grouted section reinforced with steel pipe or steel reinforcing. Micropiles develop their load-carrying capacity through soil, bedrock, or a combination of the two. The full length of the micropile must contain either a steel pipe or steel reinforcement. One of the materials used with micropiles is grout.

Grout must have a 28-day specified compressive strength no less than 4, 000 psi. As with all piles, micropiles must be reinforced. For piles or portions of piles grouted inside a temporary or permanent casing or inside a hole drilled into bedrock, the steel pipe or reinforcement must be designed to carry at least 40 percent of the design compression load.

Where a steel pipe is used for reinforcement, the portion of the cement grout enclosed within the pipe is permitted to be included at the allowable stress of the grout. The provisions for seismic reinforcement differ from these provisions. Any building or structure that is deemed to be of Seismic Design Category C must have a permanent steel casing from the top of the pile down 120 percent times the flexural length. If a building or structure is of Seismic Design Category D, E, or F, this type of pile will be considered as an alternative system.

You can use rotary or percussive drilling, with or without casing, to form a hole for the pile. The pile must be grouted using a fluid cement grout and pumped through a tremie pipe that extends to the bottom of the pile until the grout comes back up to the top. There are six requirements that must be applied to specific installation methods:

- For piles grouted inside a temporary casing, the reinforcing steel must be inserted prior to withdrawal of the casing.

- The casing must be withdrawn in a controlled manner with the grout level maintained at the top of the pile to ensure that the grout completely fills the drill hole.

- Make sure that you monitor the grout level inside the casing when you are withdrawing it to ensure that there is nothing obstructing the flow of the grout.

- You must verify the design diameter of the drill hole for a pile that is grouted in an open hole in soil without temporary casing.

- By using a suitable means for piles designed for end bearing, you will be verifying that the bearing surface is properly cleaned prior to grouting.

- Subsequent piles cannot be drilled near piles that have been grouted until the grout has had enough time to harden.

• You must grout piles as soon as possible after you have completed drilling.

• With piles designed with full-length casing, the casing must be pulled back to the top of the bond zone and reinserted to verify grout coverage outside it.

Pier Foundations

Isolated piers used as foundation must comply with minimum dimensions of 2 feet, with the height not exceeding 12 times the least horizontal dimension. Reinforcements where required must be assembled and tied together and must be placed in the pier hole as a unit before the reinforced portion of the pier is filled with concrete. This does not apply to steel dowels that have been embedded 5 feet or less in the pier. Please note this exception: reinforcement is permitted to be wet-set, and the 2 1/2 -inch concrete cover requirements can be reduced to 2 inches for Group R-3 and U occupancies that are not more than two stories and are of light-frame construction, provided that the construction method can be demonstrated to the satisfaction of the building official.

Place concrete in such a way that any foreign matter is taken out and that a full-sized shaft is secured. You may not place concrete through water unless a tremie or other method has been approved. Do not just chute the concrete directly into the pier. Concrete must be poured in a rapid and continuous operation through a funnel hopper that you have placed in the center at the top of the pier. If you find that the pier foundation has belled at the bottom, you must check to see that the edge thickness of the bell is not less than what is required for the edge of footing.

CHAPTER 19
CONCRETE

For clarification purposes in this chapter text in italics represents provisions that differ from the American Concrete Institute ACI 318 and the International Building Code requirements for structural concrete. Structural concrete must be designed and constructed in accordance with the requirements of both, and amendments in this chapter clarify the two. There are provisions in this chapter for the design and construction of slabs on grade that will not be governed by this chapter unless they send vertical loads or lateral forces from other parts of the structure to the soil. This is pointed out in the appropriate section of this chapter.

What all this means is that you must refer to ACI 318 and the provisions of this chapter to ensure that the materials, quality control, design, and construction of concrete used in structures meets the provision of all codes involved. The definitions of relevant terms are found in ACI 318; please consult it. Construction documents for projects that use structural concrete must include the following information:

• The specified compressive strength of concrete at the stated ages or stages of construction for which each element is designed

• The specified strength or grade of reinforcement

• The size and location of structural elements, reinforcement, and anchors

• Provision for dimensional changes resulting from creep, shrinkage, and temperature

• The magnitude and location of prestressing forces

• Anchorage length of reinforcement and location and length of lap splices

• Type and location of mechanical and welded splices of reinforcement

• Details and location of contraction or isolation joints specified for plain concrete

• Minimum concrete compressive strength at time of post-tensioning

• Stressing sequence for post-tensioning tendons

• For structures assigned to Seismic Design Category D, E, or F, a statement if slab on grade is designed as a structural diaphragm (see Section 21.10.3.4 of ACI 318)

Please refer to Chapter 17 for the special inspection of concrete elements of buildings and structures.

SPECIFICATIONS FOR TESTS AND MATERIALS

Materials used to produce and test concrete must comply with the applicable standards listed in ACI 318; where required, special instructions and tests must be in accordance with Chapter 17. Glas- fiber-reinforced concrete (GFRC) and the materials used in glass concrete must be in accordance with the PCI MNL (Precast Prestressed Concrete Institute) 128 standard. Note that the italics represent differing provisions.

DURABILITY REQUIREMENTS

Durability is the ability of concrete to resist weathering action and abrasion while maintaining properties. Durability requirements include water-cementitious materials; refer to ACI 318. Where maximum water-cementitious materials ratios are specified in ACI 318, they must be calculated in accordance to Section 4.1.

Freezing and thawing exposures are also a concern when working with concrete. Precautions must be taken with concrete that is going to be exposed to a possible freezing and thawing situation. Concrete that is exposed to freezing and thawing or to deicing chemicals must be air-entrained. Air entrainment is the intentional creation of tiny air bubbles in concrete. The primary purpose is to increase durability. Please refer to ACI 318, Section 4.2.1, for more detailed information.

Concrete that will be subjected to the following exposures must conform to the corresponding maximum water-cementitious materials ratios and minimum specified concrete compressive-strength requirements of ACI 318, Section 4.2.2:

• Concrete intended to have low permeability where exposed to water
• Concrete exposed to freezing and thawing in a moist condition or to deicing chemicals
• Reinforced concrete exposed to chlorides from deicing chemicals, salt, salt water, brackish water, seawater, or spray from any of these sources
• Exception: occupancies and appurtenances thereto in Group R s that are in buildings less than four stories in height utilizing normal-weight aggregate concrete must comply with the requirements of Table 19.1 based on the weathering classification (freezing and thawing) determined by Figure 19.1

other. However, these provisions do not apply to anchors that are installed in hardened concrete. Be sure that the bolts you use conform to ASTM A 307 or an equally approved reference. Table 19.2 shows the allowable service load on embedded bolts. There is also an equation that must be satisfied in relation to this table, which you will find in Equation 19.1.

$$(P_s/P_t)^{5/3} + (V_x/V_t)^{5/3} \leq 1 \qquad \text{(Equation 19.1)}$$

Table 19.2 provides information for service loads in tension for the edge distance and spacing. The code permits a reduction of 50 percent, with a reduction in allowable service load; these must be equal reductions. An increase by one-third is allowed but only according to the provisions of Chapter 16. According to Table 19.2, a 100-percent increase is allowed where provided for the installation of anchors. Please note that there is no allowable increase in shear value.

SHOTCRETE

Shotcrete is a process in which concrete or mortar is shot at high pressure onto a surface. Shotcrete must conform to the provisions of this section for reinforced or

TABLE 19.2 Allowable service load on embedded bolts (pounds).

BOLT DIAT-MER (inches)	MINI-MUM EM-BED-MENT (inches)	EDGE DIS-TANCE (inches)	SPACING (inches)	MINIMUM CONCRETE STRENGTH (psi)					
				$f'_c = 2,500$		$f'_c = 3,000$		$f'_c = 4,000$	
				Tension	Shear	Tension	Shear	Tension	Shear
$1/4$	$2^1/2$	$1^1/2$	3	200	500	200	500	200	500
$3/8$	3	$2^1/4$	$4^1/2$	500	1,100	500	1,100	500	1,100
$1/2$	4	3	6	950	1,250	950	1,250	950	1,250
	4	5	5	1,450	1,600	1,500	1,650	1,550	1,750
$5/8$	$4^1/2$	$3^3/4$	$7^1/2$	1,500	2,750	1,500	2,750	1,500	2,750
	$4^1/2$	$6^1/4$	$7^1/2$	2,125	2,950	2,200	3,000	2,400	3,050
$3/4$	5	$4^1/2$	9	2.250	3,250	2,250	3,560	2,250	3,560
	5	$7^1/2$	9	2,825	4,275	2,950	4,300	3,200	4,400
$7/8$	6	$5^1/4$	$10^1/2$	2,550	3,700	2,550	4,050	2,550	4,050
1	7	6	12	3,050	4,125	3,250	4,500	3,650	5,300
$1^1/8$	8	$6^3/4$	$13^1/2$	3,400	4,750	3,400	4,750	3,400	4,750
$1^1/4$	9	$7^1/2$	15	4,000	5,800	4,000	5,800	4,000	5,800

For SI: 1 inch = 25.4 mm, 1 pound per square inch = 0.00689 MPa, 1 pound = 4.45 N.

Tradetip

ACI 318, Section 11.11, governs special provisions for columns and contains the following provisions: Category B Seismic Design structures that have columns of ordinary moment frames must be designed for shear in accordance with 21.12.3. This applies to frames with a clear height-to-maximum-plan-dimension ratio of five or less.

Tradetip

Modification to ACI 318, Section 10.5, is made by adding a new section 10.5.5 that reads: In exterior columns of buildings that are categorized as Seismic Design B, there must be at least two main reinforcing bars continuously in the top and bottom of beams in ordinary movement frames.

Tradetip

Modifications of existing definitions and addition of new definitions to ACI 318, Section 21.1, are as follows:
  • Detailed Plain-Concrete Structural Wall: A wall that complies with ACI 318, chapter 22, and includes 22.6.7.
  • Ordinary Precast Structural Wall: A precast wall complying with Chapters 1-18 of ACI 318.
  • Ordinary Reinforced-Concrete Structural Wall: A cast-in-place wall that complies with chapters 1 through18.
  • Ordinary Structural Plain-Concrete Wall: A wall that complies with the requirements of Chapter 22 but not 22.6.7.
  • Wall Pier: Part of a wall that has a length-to-wall thickness of at least 2.5. This cannot exceed 6, with a clear height of at least two times its horizontal length.

Above all else remember that the details of connections and splices must be made only by methods that have been approved and even then they will only be approved if tested in accordance with the approved rules. A building official or an approved representative of the manufacturer will inspect concrete-filled pipe columns that are shop-fabricated.

CHAPTER 20
ALUMINUM

Aluminum used for structural purposes in buildings must comply with AA ASM 35 and AA ADM 1 and therefore there is no chapter commentary. Please refer to Chapter 16 for the nominal load requirements.

!Codealert

Autoclaved aerated concrete is a low-density cementitious product of calcium silicate hydrates whose material specifications are defined in ASTM C 1386.

QUALITY ASSURANCE

Constructed masonry must follow a quality-assurance program to ensure compliance with any construction documents that you handed in to the building official at the beginning of your project. Chapter 17 contains the inspection and testing requirements that a quality-assurance program must comply with.

Each wythe must be tested for compressive strength by using the unit-strength or the prism-test method. Let's take a look at the unit-strength method. Table 21.3 demonstrates that the compressive strength of clay masonry is determined by the strength of the units and the type of mortar specified. Table 21.4 shows the compressive strength of concrete masonry. You can't go by these tables alone. You must make sure that clay units conform to ASTM C 62, ASTM C 216, or ASTM C 652, that you have sampled your units, and that they are tested in accordance with ASTM C 67.

Concrete units have to conform to ASTM C 55 or ASTM C90 and are sampled and tested in accordance with ASTM C 140. Both clay and concrete bed joints may not be more than 5/8 inch thick, conform to ASTM C 476, and have a compressive strength that is no less that 2,000 psi. ASTM C 1019 contains provisions for the compressive strength of grout.

As you can see, the unit-strength method verifies the compressive strength of the individual materials and then uses tables to determine the compressive strength of the assembly. The procedure to determine the compressive strength of AAC masonry is similar and contains some of the same provisions as clay and concrete masonry, except that it is based on the strength of the AAC masonry unit only, units must also conform to ASTM C 1386, and the thickness of bed joints cannot be more than 1/8 inch.

Where tables are not used, the prism test method comes in. In the prism-test method you construct prisms at the job site to verify compliance with design compressive strength. Prism testing is done when specified in your construction documents and when masonry does not meet the requirements of the strength testing.

!Codealert

A foundation pier is an isolated vertical foundation member whose horizontal dimension measured at right angles to its thickness does not exceed three times its thickness and whose height is equal to or less than four times its thickness.

TABLE 21.3 Compressive strength of clay masonry.

NET AREA COMPRESSIVE STRENGTH OF CLAY MANSONRY UNITS (psi)		NET AREA COMPRESSIVE STRENGTH OF MASONRY (psi)
Type M or S mortar	Type N mortar	
1,700	2,100	1,000
3,350	4,150	1,500
4,950	6,200	2,000
6,600	8,250	2,500
8,250	10,300	3,000
9,900	—	3,500
13,200	—	4,000

For SI: 1 pound per square inch = 0.00689 MPa.

You must construct three prisms, which will be tested in accordance with ASTM C 1324. The building official must be consulted and give permission before such testing is allowed. The masonry prisms must be at least 28 days old before using them for testing and must be cut (by saw) from masonry of at least 5,000 square feet of the area in question. You must follow the guidelines of ASTM C 1324 when conducting prism tests; this is also true for the transportation and preparation of prisms.

SEISMIC DESIGN

This section of the 2006 International Building Code must also be in compliance with ACI 530/ASCE 5 or TMS 402, depending on the seismic structure category of your building, as determined in chapter 16.

Unless isolated on three edges from in-plane motion (of the basic structural systems), masonry walls are considered to be part of the seismic-force-resisting system. Please refer to ACI 530/ASCE 5 and TMS 402 for the provisions relating to seismic design of your building or structure.

TABLE 21.4 Compressive strength of concrete masonry.

NET AREA COMPRESSIVE STRENGTH OF CONCRETE MANSONRY UNITS (psi)		NET AREA COMPRESSIVE STRENGTH OF MASONRY (psi)
Type M or S mortar	Type N mortar	
1,250	1,300	1,000
1,900	2,150	1,500
2,800	3,050	2,000
3,750	4,050	2,500
4,800	5,250	3,000

For SI: 1 inch = 25.4 mm, 1 pound per square inch = 0.00689 MPa.
a. For units less than 4 inches in height, 85 percent of the values listed.

!Codealert
Masonry in which the tensile resistance is taken into consideration and the resistance of the reinforcing steel, if present, is neglected is known as unreinforced masonry.

EMPIRICAL DESIGN OF MASONRY

Empirical design of masonry is to conform to this section or chapter 5 of ACI 530/ASCE 5/TMS 420. There are limitations on the use of empirical design masonry. You may not use empirical design for the following:

- Buildings of Seismic Design Category D, E, or F or the seismic-force-resisting system in buildings of Seismic Design Category B or C

- Masonry elements that are part of the lateral-force-resisting systems with wind speeds over 110 mph

- Interior masonry elements that are not part of the lateral-force-resisting system in buildings other than those specified in Chapter 6 of ASCE 7.

- Outside masonry elements not part of the lateral-force-resisting system that are more than 35 feet above ground or elements that are equal to or less than 35 feet above ground where the wind speed is over 110 mph

- AAC masonry

Structures that rely on walls for lateral support must have shear walls to provide parallel support. Such shear walls have to be positioned in two separate planes. Shear walls have to have a cumulative length of at least 0.4 times the long dimension of the building. The cumulative length of a shear wall is not to include openings or any other element with a length that is less than one-half its height. Refer to Table 21.5 for the provisions for diaphragm length-to-width ratios.

Be sure to comply with the provisions of the code for dry-stacked, surface-bonded concrete-masonry walls. For instance, such walls must be of adequate strength as listed in Table 21.6, however, any allowable stresses that are not specified in this table must comply with the requirements of ACI 530/ASCE 5/TMS 402.

Please refer to Table 21.7 for the allowable compressive stresses for empirical design of masonry.

The following list contains the minimum thickness of masonry walls:

- Masonry bearing walls that are more than one story high must be at least 8 inches thick but no less than 6 inches for bearing walls of one-story buildings.

!Codealert

Thin-bed mortar is mortar for use in construction of AAC unit masonry with joints 0.06 inch or less.

!Definitionalert

Rubble Stone Masonry: Masonry composed of roughly shaped stones.

TABLE 21.5 Diaphragm length-to-width ratios.

FLOOR OR ROOF DIAPHRAGM CONSTRUCTION	MAXIMUMLENGTH-TO-WIDTH RATIO OF DIAPHRAGM PANEL
Cast-in-place concrete	5:1
Precast concrete	4:1
Metal deck with concrete fill	3:1
Metal deck with no fill	2:1
Wood	2:1

TABLE 21.6 Allowable stress gross cross-sectional area for dry-stacked, surface-bonded concrete masonry walls.

DESCRIPTION	MAXIMUM ALLOWABLE STRESS (psi)
Compression standard block	45

Flexural tension
  Horizontal span
  Vertical span
	30
	18

Shear	10

For SI: 1 pound per square inch = 0.00689 MPa.

!Codealert

See Section 2103 for current code requirements on mortar for AAC masonry.

• Rough, random, or coursed rubble walls must have a minimum thickness of 16 inches.

• Shear walls and foundation piers must have a minimum thickness of 8 inches.

• Parapet walls must have a minimum thickness of 8 inches; however their height cannot be three times their thickness. Chapter 15, Sections 1503.2 and 1503.3, contain additional provisions for parapet walls that you must follow.

Although foundation walls must have a minimum thickness of 8 inches, I didn't include them in the list above because they also must be in compliance with Table 21.8 for foundation-wall construction and the following conditions:

• The foundation wall cannot be higher than 8 feet between lateral supports.

• The environment that surrounds the foundation must be graded to drain surface water away from foundation walls

• All backfill must be properly drained for the removal of ground water from foundation walls

• Lateral support must be provided to the top of the foundation before backfilling

• The foundation walls must have a maximum of three times the basement-wall height between the masonry walls.

• Nonexpansive soil areas must exist and backfill must be granular.

• Type M or S mortar must be used for masonry laid in running bond.

If the above requirements for foundation walls are not met, provisions for foundation walls must be designed in accordance with chapter 18, Section 1805.5, regarding footings and foundations.

Multiwythe Masonry Walls

Multiwythe masonry walls must be bonded, both facing and backing, in accordance with provisions for bonding with masonry headers, wall ties, joint reinforcements, or natural or cast stone. Bonding with masonry headers includes solid units where no less than 4 percent of the wall surface of each face can be made of headers that extend longer than 2 inches into the backing. There may not be a distance of more than 24 inches between each bordering full-length header.

If there are walls in which there is a single header that doesn't go through the wall, headers from opposite sides have to overlap by 3 inches. Bonding with masonry headers includes hollow units as well, and if you've used two or more units to make up the thickness of one wall, you must bond stretcher courses at vertical intervals. These cannot be more than 34 inches or lap over the unit below by more than 3 inches.

The last component of bonding with masonry headers is masonry-bonded hollow walls. In these walls, the facing and backing are bonded to prevent less than 4 percent of the wall surface of each face is made of masonry-bonded units that do not extend more than 3 inches into the backing. Bonding with wall ties or joint re-

TABLE 21.7 Allowable compressive stresses for empirical design of masonry.

CONSTRUCTION: COMPRESSIVE STRENGTH OF UNIT GROSS AREA (psi)	ALLOWABLE COMPRESSIVE STRESSES[a] GROSS CROSS-SECTIONAL AREA (psi)	
	Type M or S mortar	Type N mortar
Masonry of brick and other solid units of clay or sand-lime or concrete brick:		
8,000 or greater	350	300
4,500	225	200
2,500	160	140
1,500	115	100
Grouted masonry, of clay or shale; sand-lime or concrete:		
4,500 or greater	225	200
2,500	160	140
1,500	115	100
Solid masonry of solid concrete masonry units:		
3,000 or greater	225	200
2,000	160	140
1,200	115	100
Masonry of hollow load-bearing units:		
2,000 or greater	140	120
1,500	115	100
1,000	75	70
700	60	55
Hollow walls (noncomosite masonry bonded)[b] Solid units:		
2,500 or greater	160	140
1,500	115	100
Hollow units	75	70
Stone ashlar masonry:		
Granite	720	640
Limestone or marble	450	400
Sandstone or cast stone	360	320
Rubble stone masonry Coursed, rough or random	120	100

For SI: 1 pound per square inch = 0.00689 MPa.

a. Linear interpolation for determining allowable stresses for masonry units having compressive strengths which are intermediate between those given in the table is permitted.

b. Where floor and roof loads are carried upon one wythe, the gross-cross-sectional area is that of the wythe under load; if both wythes are loaded, the gross cross-sectional area is that of the wall minus the area of the cavity between the wythes. Walls bonded with metal ties shall be considered as noncomposite walls unless collar joints are filled with mortar or grout.

TABLE 21.8 Foundation wall construction.

WALL CONSTRUCTION	NOMINAL WALL THICKNESS (inches)	MAXIMUM DEPTH OF UNBALANCED BACKFILL (feet)
Fully grouted masonry	8	7
	10	8
	12	8
Hollow unit masonry	8	5
	10	6
	12	7
Solid unit masonry	8	5
	10	7
	12	7

For SI: 1 inch = 25.4 mm, 1 foot = 304.8 mm.

inforcements includes bonding with wall ties, adjustable ties, and prefabricated joint reinforcement. First, let's take a look at the provisions for wall ties or joint reinforcements.

A maximum vertical distance of 24 inches is mandatory between ties, with a maximum horizontal distance of 36 inches. The rods or ties have to be bent in a rectangular shape, and the hollow-masonry units are laid on the vertical with the cells. Bonding with wall ties requires the facing and backing of masonry walls to be bonded with a wire size of W2.8 or of a metal wire of equal size.

Bonding with adjustable wall ties does not have the same provisions as bonding with wall ties. In bonding with adjustable wall ties there must be at least one tie for each 1.77 square feet of wall area. The spacing of adjustable wall ties cannot be more than 16 inches either vertically or horizontally. The bed joints must have a maximum vertical offset from one wythe to the other of 1 1/4 inches. The connecting parts of the ties must have a maximum clearance of 1/16 inch. Ties have to have at least two wires of size W2.8 when pintle legs are used.

Bonding with prefabricated joint reinforcement must have at least one cross wire for each 2 2/3 square feet of wall area. Vertical spacing of joint reinforcing cannot be more than 24 inches, and cross wires on prefabricated joint reinforcing must not be less than W1.7 in size.

The second part of bonding is bonding with natural or cast stone. This includes ashlar and rubble stone masonry.

In ashlar masonry, uniformly distributed bonder units must be provided and must not be less than 10 percent of the wall area. Bonder units must not extend less

!Definitionalert

Ashlar Masonry: Masonry made of various-sized rectangular units that have sawn, dressed, or squared bed surfaces and that have been properly bonded and laid in mortar.

than 4 inches into the backing wall. There must be bonder units with a maximum spacing of 36 inches vertically and horizontally for every rubble stone masonry 24 inches or less in thickness. If the masonry is more than 24 inches thick, there must be a bonder unit for each 6 square feet of wall surface on both sides.

Masonry walls that depend on each other for lateral support must be anchored with a metal rod, wire, or strap. They must be anchored in places where they meet or by one of the methods listed below:

- Half of the units may be laid in an overlapping masonry bonding pattern.
- Alternate units have a bearing not less than 3 inches on the unit below.
- Steel connectors are used to anchor walls. Anchors must be at least 24 inches long with a maximum of 48-inch spacing.
- Joint reinforcement used to anchor walls must be spaced a maximum of 8 inches.
- Wires for such use must be at least size W1.7 and extend for at least 30 inches in each direction at the intersection.
- Interior nonloading walls must be anchored at vertical intervals of their intersections of not more than 16 inches. Use joint reinforcement or 1/4-inch-mesh galvanized hardware cloth.
- Ties, joint reinforcement, or anchors, if used, must be spaced so that an equivalent area of anchorage is provided in accordance with this section.

!Codealert

Sections 2104 and 2105 deserve your attention for changes in code requirements.

!Codealert

Investigate code changes in Section 2107 for allowable stress design.

GLASS-UNIT MASONRY

Glass-unit masonry is masonry composed of glass units held by mortar. This section covers the empirical requirements for nonload glass-unit masonry elements for exterior or interior walls. There are limitations on the use of glass masonry. For instance, neither solid nor hollow glass block can be used in fire walls, party walls, fire barriers, or fire partitions. These types of block have to be built with mortar, and you must use reinforcement by structural frames, masonry, or concrete recesses. There are two exceptions to this. One is that glass-block assemblies that have a fire-protection rating no less than 2/4 hour will be allowed as opening protectives. (See chapter 7 for more information.) The second exception is for glass-block assemblies permitted in chapter 4, Section 404.5. If you are using solid or hollow glass-block units, please be aware that they must be standard or thin units.

A standard unit has a specified thickness of 3 7/8 inches, and a thin unit has a specified thickness of 3 1/8 inches for hollow and a minimum of 3 inches for solid units. There are five types of panel size; please consult the following list for regulations of these panels:

• Exterior standard-unit panels: Each individual exterior panel must be 144 square feet when the wind pressure is 20 psf. There must be 25 feet in width or 10 feet in height between structural supports. Panel areas are allowed to be adjusted; please see Figure 21.1 for details.

• Exterior thin-unit panels: These panels must have a maximum area of 85 square feet with a maximum dimension of 15 feet in width or 10 feet in height between structural supports.

• Interior panels: Interior panels must have a maximum area of 250 square feet for a standard unit and a maximum of 150 feet for thin-unit panels.

• Solid units: These panels must have an area of solid glass-block panel in both inside and outside walls of not more than 100 square feet.

All glass-unit masonry panels must be isolated so that in-plane loads are not imparted to the panel. Installed glass-unit masonry that has an installed weight of 40 psf or less with a maximum height of 12 feet is allowed to be supported on

TABLE 21.9 Net cross-sectional area of round flue sizes[a].

FLUE SIZE, INSIDE DIAMETER (inches)	CROSS-SECTIONAL AREA (square inches)
6	28
7	38
8	50
10	78
10 $^3/_4$	90
12	113
15	176
18	254

For SI: 1 inch = 25.4 mm, 1 square inch = 645.16 mm².
a. Flue sizes are based on ASTM C 315.

!Code alert

See Section 2112 for masonry heaters to come up to speed with current code requirements.

> **!Code**alert
>
> See Section 2113.16.2 for net cross-sectional area of square and rectangular flue size requirements.

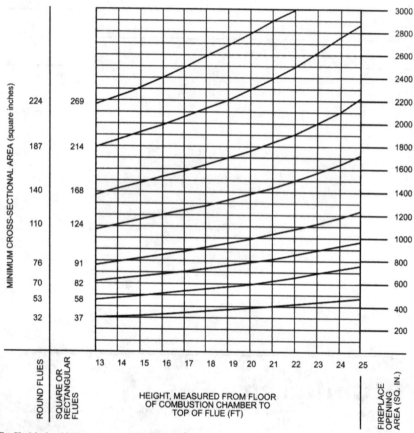

For SI: 1 inch = 25.4 mm, 1 square inch = 645 mm².

FIGURE 21.4 Flue sizes for masonry chimneys.

TABLE 21.10 Net cross-sectional area of square an rectangular flue sizes.

FLUE SIZE, INSIDE DIAMETER (inches)	CROSS-SECTIONAL AREA (square inches)
4.5 × 8.5	23
4.5 × 13	34
8 × 8	42
8.5 × 8.5	49
8 × 12	67
8.5 × 13	76
12 × 12	102
8.5 × 18	101
13 × 13	127
12 × 16	131
13 × 18	173
16 × 16	181
16 × 20	222
18 × 18	233
20 × 20	298
20 × 24	335
24 × 24	431

For SI: 1 inch = 25.4 mm, 1 square inch = 645.16 mm^2.

CHAPTER 22

STEEL

Chapter 22 gives the provisions for the design, fabrication, and erection of steel buildings and structures. Cold-formed steel construction is made up in part or entirely of steel members cold-formed to shape from sheet or strip steel. A steel joist is any steel structural member of a building that is made of hot-rolled or cold-formed solid or open-web sections. A structural-steel member is any steel member of a building that consists of a rolled-steel structural shape other than cold-formed steel or steel joist members.

IDENTIFICATION AND PROTECTION

Steel must conform to ASTM standards or to other specifications and to the provisions of this chapter. You must test any steel that is not identified as to grade to determine conformity to such standards. Steel is often painted to add a different look, but even the painting of structural steel needs to comply with requirements. You will find such requirements in AISC 360. All steel, except where fabricated of approved corrosion-resistant coating, must be protected against corrosion with a coat of approved paint or enamel.

STRUCTURAL STEEL

It is a requirement of the 2006 International Building Code that the design, fabrication, and erection of structural steel be in accordance with AISC 360. And if necessary, refer to Chapter 18 for the appropriate Seismic Design Category. See the following list for the appropriate standards:

- Seismic Design Category A, B, or C provisions are found in Section 12.2.1 of ASCE 7.
- Seismic Design Category D, E, or F provisions are found in AISC 341, Part I.

For design, construction, and quality of composite steel and concrete components, see the requirements of the AISC 360 and ACI 318. An R factor as set forth in Section 12.2.1 of ASCE 7 for the appropriate composite steel is permitted where the structure is designed and detailed in accordance with the provisions of AISC 341, Part II. The design for Seismic Design Category B or above must conform to the requirements of AISC 341, Part II. You are allowed to have composite structures in Seismic Design Categories D, E, and F, but they are subject to the provisions in 12.2.1 of ASCE 7, where evidence is provided to show that the system will perform in accordance with AISC 341, Part II.

STEEL JOISTS

The Steel Joist Institute (SJI) has specifications that the design, manufacture, and use of open-web-steel joist girders must adhere to. They are SJI K-1.1, SJI LH/DLH-1.1, and SJI JG-1.1. All construction documents written by the registered design professional must indicate the steel-joist and/or steel-joist-girder description from the above specifications. Documents must also include the requirements for joist and joist-girder design, end supports, and bearing connections. These documents must indicate requirements for special loads, which must include the following:

• Concentrated loads

• Nonuniform loads

• Net uplift loads

• Axial loads

• End moments

• Connection forces

Documents must also include special considerations that include the following:

• Joist-girder configurations and profiles for nonstandard joists

• Oversized or nonstandard web openings

• Extended ends

The last element that is required in documentation is deflection criteria for live and total loads for non-SJI standard joists. Included in the provisions for steel joists are calculations. It is up to the manufacturer to design the steel joists and/or girders in accordance with the current SJI specifications. It's possible that a registered design professional may want the joist and joist-girder calculations to be provided by the manufacturer.

When the design calculations are received, they must include a cover letter that has the seal and signature of the joist manufacturer's registered design professional. Also included in this package, still under seal and signature, are the following:

• Non-SJI standard details of bridging, such as net uplift, cantilevered conditions, etc.

• Connection details for non-SJI standard connections

• Field splices

• Joist headers

You must provide steel-joist-placement plans to show the steel joist products as they are specified on construction documents. These plans will be used for field installation. Steel-placement plans must include information such as the following:

• The listings of all applicable loads as stated in Section 2206.2 and used in the design as specified in the construction documents

• Profiles for all nonstandard joist and joist-girder configurations

• Connection requirements for:

- Joist supports

- Joist-girder supports

- Field splices

- Bridging attachments

• Live and total load-deflection criteria for non-SJI standard joints

• Connections, location, and size for all bridging

• Joist headers

The seal and signature of the joist manufacturer's registered design professional is not needed for steel-joist-replacement plans. The steel-joist manufacturer must submit a certificate of compliance when fabrication is complete. This states that the work was performed in accordance with approved construction documents and with SJI specifications.

STEEL CABLE STRUCTURES

Please refer to ASCE 19 for the design, fabrication, and erection of steel-cable structures, including connections and protective coatings of steel cables. ASCE

!Codealert
See Section 2206 for changes in code regarding steel joists.

19 also contains the provisions for the design strength of steel cables. A load factor of 1.1 must be applied to the prestress force as defined in Section 3.12 of the IBC, and in Section 3.2.1 you must replace item c with "1.5 T3" and Item d with "1.5 T4."

STEEL STORAGE RACKS

Please refer to RMI Specifications for the Design, Testing, and Utilization of Industrial-Steel Storage Racks for the design, testing, and utilization of industrial-steel storage racks and the provisions thereof. Where required, you must use Section 15.5.3 of ASCE 7 for the seismic design of storage racks.

COLD-FORMED STEEL

Please see AISI-NAS for provisions for the design of cold-formed carbon and low-alloy-steel structural members. Such design must be in accordance with ASCE 8. Cold-formed-steel light-framed design, construction, and installation must be in accordance with AISI-General and AISI-NAS. The following list contains the appropriate standards:

- Headers: Cold-formed-steel box headers, back-to-back headers, and single and double L-headers must be in accordance with AISI-Header and are subject to the limitations within.

- Trusses: Please refer to AISI-Truss for the provisions for the design, quality assurance, and testing of cold-formed-steel trusses.

- Wall stud design: Cold-formed-steel studs used for structural and nonstructural walls must be in accordance with AISI-WSD.

- Lateral design: AISI-Lateral contains the provisions for the design of light-framed cold-steel walls and diaphragms to resist wind and seismic loads.

- Prescriptive framing: Detached one- and two-family dwellings and townhouses (up to two stories) are allowed if constructed in accordance with AISI-PM and subject to the limitations therein.

CHAPTER 23
WOOD

This chapter contains provisions on materials, design, and construction of wood and wood fasteners. I have added definition alerts in appropriate sections, and you will find tables and figures throughout.

There has been a change in the code regarding the general design requirements of structural elements or systems that are constructed in part or in whole of wood or wood-based products, which you will find in the following list:

- Sections 2304, 2305, and 2306 contain the requirements for allowable-stress design.
- Sections 2304, 2305, and 2307 contain the requirements for load and resistance-factor design.
- Sections 2304 and 2308 contain the requirements for conventional light-frame construction.
- Note this exception: Buildings that have been designed in accordance with the provisions of the AF&PA WFCM are considered to have met the requirements of Section 2308.

Lumber is presumed to be measured in nominal dimensions unless actual dimensions are noted.

MINIMUM STANDARDS AND QUALITY

This section contains the minimum standards and quality for the different types of wood and other materials used in construction, including but not limited to: preservative-treated wood, structural log members, prefabricated wood I-joists, staples, and nails.

Sawn lumber that is used for load-supporting purposes must be identified by the mark of a lumber-grading agency. You can use an inspection agency too, but keep in mind that the agency you use must be approved by an accreditation body that complies with DOC PS 20 or an equivalent. All grading practices and identification must comply with the procedures of DOC PS 20 as well. A certificate of inspection may be accepted instead of the grade mark but only if all provisions have been met. You are also permitted to use approved end-jointed lumber instead of solid-sawn members as long as they are of the same species and grade.

There are many elements in this chapter that must also follow the guidelines set forth in other referenced standards. Some of these include the following:

- Prefabricated wood I-joists—ASTM D 5055
- Structural glued-laminated timber—AITC A 190.1 and ASTM D 3737
- Wood structural panels—DOC PS 1 or PS 2
- Hardboard—AHA A 135.6
- Prefinished hardboard—AHA A 135.5
- Particleboard—ANSI A208.1
- Preservative-treated wood—AWPA U1 and M4

Fiberboard

Fiberboard and its various uses must conform to ASTM C 208 as well as fiberboard-sheathing standards. When working with fiberboard, don't forget jointing, roof and wall insulation, and protection. Edges must be tight-fitting with square or U-shaped joints. When fiberboard is used as insulation, regardless of construction type, you must use an approved roof covering.

Fiberboards are allowed as wall insulation in all types of construction as long as they are installed and fireblocked to comply with Chapter 7. If used in fire walls

---

!Codealert

The design of structural elements or systems constructed partially or wholly of wood or wood-based products must be in accordance with one of the following methods:
- Allowable stress design in accordance with Sections 2304, 2305 and 2306
- Load and resistance factor design in accordance with Sections 2304, 2305 and 2307
- Conventional light-frame construction in accordance with Sections 2304 and 2308.

TABLE 23.1 Allowable spans and loads for wood structural panel sheathing and single-floor grades continuous over two or more spans with strength axis perpendicular to supports.[a,b] *(continued)*

SINGLE FLOOR GRADES		ROOF[c]				FLOOR[d]
		Maximum span (inches)		Load[e] (psf)		Maximum span
Panel span rating	Panel thickness (inches)	With edge support[f]	Without edge support	Total load	Live load	(inches)
16 o.c.	$1/2$, $19/32$, $5/8$	24	24	50	40	16[h]
20 o.c.	$19/32$, $5/8$, $3/4$	32	32	40	30	20[h,j]
24 o.c.	$23/32$, $3/4$	48	36	35	25	24
32 o.c.	$7/8$, 1	48	40	50	40	32
48 o.c.	$1^3/32$, $1^1/8$	60	48	50	40	48

For SI: 1 inch = 25.4 mm, 1 pound per foot = 14.59 N/m.

a. Applies to panels 24 inches or wider.

b. Floor and roof sheathing conforming with this table shall be deemed to meet the design criteria of Section 2304.7.

c. Uniform load deflection limitations 1/180 of span under live load plus dead load, 1/240 under live load only.

d. Panel edges shall have approved tongue-and-groove joints or shall be supported with blocking unless 1/4-inch minimum thickness underlayment or 1 1/2 inches of approved cellular or lightweight concrete is placed over the subfloor, or finish floor is 3/4-inch wood strip. Allowable uniform load based on deflection of 1/360 of span is 100 pounds per square foot except the span rating of 48 inches on center is based on a total load of 65 pounds per square foot.

e. Allowable load at maximum span.

f. Tongue-and-groove edges, panel edge clips (one midway between each support, except two equally spaced between supports 48 inches on center), lumber blocking or other. Only lumber blocking shall satisfy blocked diaphragm requirements.

g. For 1/2-inch panel, maximum span shall be 24 inches.

h. Span is permitted to be 24 inches on center where 3/4-inch wood strip flooring is installed at right angles to joist.

i. Span is permitted to be 24 inches on center for floors where 1 1/2 inches of cellular or lightweight concrete is applied over the panels.

TABLE 23.2 Allowable span for wood structural panel combination subfloor-underlayment (single floor).[a,b] (Panels continuous over two or more spans and strength axis perpendicular to supports.)

IDENTIFICATION	MAXIMUM SPACING OF JOISTS (Inches)				
	16	20	24	32	48
Species group[c]	Thickness (Inches)				
1	$^1/_2$	$^5/_8$	$^3/_4$	—	—
2, 3	$^5/_8$	$^3/_4$	$^7/_8$	—	—
4	$^3/_4$	$^7/_8$	1	—	—
Single floor span rating[d]	16 o.c.	20 o.c.	24 o.c.	32 o.c.	48 o.c.

For SI: 1 inch = 25.4 mm, 1 pound per square inch = 0.0479 kN/m².

a. Spans limited to value shown because of possible effects of concentrated loads. Allowable uniform loads based on deflection of 1/300 of span is 100 pounds per square foot except allowable total uniform load for 1 $^1/_8$-inch wood structural panels over joists spaced 48 inches on center is 65 pounds per square foot. Panel edges shall have approved tongue-and-groove joints or shall be supported with blocking, unless 1/4-inch minimum thickness underlayment or 1 $^1/_2$ inches of approved cellular or lightweight concrete is place over the subfloor, or finish floor is $^3/_4$-inch wood strip.

b. Floor panels conforming with this table shall be deemed to meet the design criteria of Section 2304.7.

c. Applicable to all grades of sanded exterior -type plywood. See DOC PS 1 for plywood species groups.

d. Applicable to Underlayment grade, C-C (Plugged) plywood, and Single Floor grade wood structural panels.

As a reminder, fastenings for wood foundations must be as required in AF&PA Technical Report No.7.

Decay and Termites

Most of us are familiar with the effect of decay on wood, but termites can have the same effect. When termites are out looking for a nice place to lunch, make sure you've taken proper precautions so that they aren't tempted to dine on your house, building, or structure. Protection against termites and decay are both important because a lot of this damage is hidden and may be too late for repairs by the time you find out. This is why wood should be naturally durable or preservative-treated.

Termite-resistant wood must be used for wood floor framing, especially in areas heavily populated with termites. Water-borne preservatives, in accordance with AWPA U1, are used to preserve such wood for use above the ground in wood joists, sleepers and sills, siding and girder ends, which are connected in one way or another to the foundation.

TABLE 23.4 Fastening schedule. *(continued)*

CONNECTION	FASTENING[a,m]		LOCATION
31. Wood structural panels and particleboard[b] Subfloor, roof and wall sheathing (to framing)	$^1/_2$" and less	6d[c,l] $2^3/_8$" $\times$ 0.113" nail[n] $1^3/_4$" $\times$ 16 gage[o]	
	$^{19}/_{32}$" to $^3/_4$"	8d[d] or 6d[e] $2^3/_8$" $\times$ 0.113" nail[n] 2" 16 gage[p]	
	$^7/_8$" to 1"	8d[e]	
Single Floor (combination sub-floor-underlayment to framing)	$1^1/_8$" to $1^1/_4$"	10d[d] or 8d[d]	
	$^3/_4$" and less	6d[c]	
	$^7/_8$" to 1"	8d[e]	
	$1^1/_8$" to $1^1/_4$"	10d[d] or 8d[d]	
32. Panel siding (no framing)	$^1/_2$" and less	6d[f]	
	$^5/_8$" and less	8d[f]	
33. Fiberboard sheathing[g]	$^1/_2$"	No. 11 gage roofing nail[b] 6d common nail (2" $\times$ 0.113") No. 16 gage staple[i]	
	$^{25}/_{32}$"	No. 11 gage roofing nail[b] 8d common nail ($2^1/_2$" $\times$ 0.131") No. 16 gage staple[i]	
34. Interior paneling	$^1/_4$"	4d[j]	
	$^3/_8$"	6d[k]	

For SI: 1 inch = 25.4 mm.
a. Common or box nails are permitted to be used except where otherwise stated.
b. Nails spaced at 6 inches on center at edges, 12 inches at intermediate supports except 6 inches at supports where spans are 48 inches or more. for nailing of wood structural panel and particleboard diaphragms and shear walls, refer to Section 2305. Nails for wall sheathing are permitted to be common, box or casing.
c. Common or deformed shank (6d-2" $\times$ 0.113"; 8d-$2^1/_2$" $\times$ 0.131"; 10d-3" $\times$ 0.148").
d. Common (6d-2" $\times$ 0.113"; 8d-$2^1/_2$" $\times$ 0.131"; 10d-3" $\times$ 0.148").
e. Deformed shank (6d-2" $\times$ 0.113"; 8d-$2^1/_2$" $\times$ 0.131"; 10d-3" $\times$ 0.148").
f. Corrosion-resistant siding (6d-$1^7/_8$" $\times$ 0.106"; 8d-$2^3/_8$" $\times$ 0.128") or casing (6d-2" $\times$ 0.099"; 8d-$2^1/_2$" $\times$ 0.113") nail.
g. Fasteners spaced 3 inches on center at exterior edges and 6 inches on center at intermediate supports, when used as structural sheathing. Spacing shall be 6 inches on center on the edges and 12 inches on center at intermediate supports for nonstructural applications.
h. Corrosion-resistant roofing nails with $^7/_{16}$-inch-diameter head and $1^1/_2$-inch length for $^1/_2$-inch sheathing and $1^3/_4$-inch length for $^{25}/_{32}$-inch sheathing.
i. Corrosion-resistant staples with normal $^7/_{16}$-inch crown and $1^1/_8$-inch length for $^1/_2$-inch sheathing and $1^1/_2$-inch length for $^{25}/_{32}$-inch sheathing. Panel supports at 16 inches (20 inches if strength axis in the long direction of the panel, unless otherwise marked).
j. Casing ($1^1/_2$" $\times$ 0.080") or finish ($1^1/_2$" $\times$ 0.072") nails spaced 6 inches on panel edges, 12 inches at intermediate supports.
k. Panel supports at 24 inches. Casing or finish nails spaced 6 inches on panel edges, 12 inches at intermediate supports.
l. For roof sheathing applications, 8d nails ($2^1/_2$" $\times$ 0.113") are the minimum required for wood structural panels.
m. Staples shall have a minimum crown width of $^7/_{16}$ inch.
n. For roof sheathing applications, fasteners spaced 4 inches on center at edges, 8 inches at intermediate supports.
o. Fasteners spaced 4 inches on center at edges, 8 inches at intermediate supports for subfloor and wall sheathing and 3 inches on center at edges, 6 inches at intermediate supports for roof sheathing.
p. Fasteners spaced 4 inches on center at edges, 8 inches at intermediate supports.

calculating lateral load resistance to transfer lateral earthquake forces in excess of 150 pounds per foot.

This section also includes provisions for the design of wood diaphragms, which can be used to resist horizontal forces but only if the deflection of the plane of the diaphragm does not exceed the permissible deflection or resisting elements. Equation 23.1 is used to calculate the deflection of a blocked wood-panel diaphragm. Tables 23.5 through 23.7 are used for calculating diaphragm dimensions.

For structures of Seismic Design Category F you must follow additional requirements. Wood structural-panel sheathing used for diaphragms and shear walls that are part of the seismic-force-resisting systems must be applied directly to the framing members, except that wood structural-panel sheathing in a diaphragm is allowed to be fastened over solid lumber planking if the panel joints and lumber planking joints do not coincide.

When designing structures with rigid diaphragms, you must refer to the requirements of Section 12.3.2 of ASCE 7 and the horizontal-shear-distribution requirements of Section 12.8.4 of ASCE 7. Figure 23.1 shows the diaphragm length and width for plan views of open-front buildings, and Figure 23.2 contains the diaphragm length and width for plan views of cantilevered diaphragms.

ALLOWABLE STRESS DESIGN

The following list contains the provisions with which the allowable stress design of wood elements must comply:

• American Forest and Paper Association
• American Institute of Timber Construction
• American Society of Agricultural Engineers
• Engineered Wood Association (formerly American Plywood Association)
• Truss Plate Institute, Inc.

Included in allowable stress design is lumber decking. Table 23.8 contains the flexure and deflection formulas prescribed for lumber decking.

There has been a change in the code for some parts of this section, such as wood diaphragms. By looking at Table 23.9 or 23.10 you will find that wood structural-panel diaphragms are allowed to resist horizontal forces by using the allowable shear capacities. The allowable capacities that are found in these two tables must be increased by 40 percent for wind design, and sheathed lumber (diagonal) diaphragms must be nailed in accordance with Table 23.9.

For the application of gypsum board or lath and plaster to wood framing, there are several steps to take to ensure proper adherence. When joint staggering, make sure the end joints of adjacent courses of gypsum board do not occur over the same stud.

TABLE 23.5 e^n Values (inches) for use in calculating diaphragm deflection due to fastener slip (Structural I).[a,d]

LOAD PER FASTENER[c] 220 (pounds)	FASTENER DESIGNATIONS[b]			
	6d	8d	10d	14-Ga staple x 2 inches long
60	0.01	0.00	0.00	0.011
80	0.02	0.01	0.01	0.018
100	0.03	0.01	0.01	0.028
120	0.04	0.02	0.01	0.04
140	0.06	0.03	0.02	0.053
160	0.10	0.04	0.02	0.068
180	—	0.05	0.03	—
200	—	0.07	0.47	—
220	—	0.09	0.06	—
240	—	—	0.07	—

For SI: 1 inch = 25.4 mm, 1 foot = 304.8 mm, 1 pound = 4.448 N.
a. Increase e_n values 20 percent for plywood grades other than Structural I.
b. Nail values apply to common wire nails or staples identified.
c. Load per fastener = maximum shear per foot divided by the number of fasteners per foot at interior panel edges.
d. Decrease e_n values 50 percent for seasoned lumber (moisture content <19 percent).

TABLE 23.6 Maximum diaphragm dimension ratios horizontal and sloped diaphragm.

TYPE	MAXIMUM LENGTH-WIDTH RATIO
Wood structural panel, nailed all edges	4:1
Wood structural panel, blocking omitted at intermediate joints	3:1
Diagonal sheathing, single	3:1
Diagonal sheathing, double	4:1

TABLE 23.7 Values of G_t for use in calculating deflection of wood structural panel shear walls and diaphragms.

VALUES OF G_t (lb/in. panel depth or width)

PANEL TYPE	SPAN RATING	OTHER				STRUCTURAL I			
		3-ply Plywood	4-ply Plywood	5-ply Plywood[a]	OSB	3-ply Plywood	4-ply Plywood	5-ply Plywood[a]	OSB
Sheathing	24/0	25,000	32,500	37,500	77,500	32,500	42,500	41,500	77,500
	24/16	27,000	35,000	40,500	83,500	35,000	45,500	44,500	83,500
	32/16	27,000	35,000	40,500	83,500	35,000	45,500	44,500	83,500
	40/20	28,000	37,000	43,000	88,500	37,000	48,000	47,500	88,500
	48/24	31,000	40,500	46,500	96,000	40,500	52,500	51,000	96,000
Single Floor	16 o.c.	27,000	35,000	40,500	83,500	35,000	45,500	44,500	83,500
	20 o.c.	28,000	36,500	42,000	87,000	36,500	47,500	46,000	87,000
	24 o.c.	30,000	39,000	45,000	93,000	39,000	50,500	49,500	93,000
	32 o.c.	36,000	47,000	54,000	110,000	47,000	61,000	59,500	110,000
	48 o.c.	50,500	65,500	76,000	155,000	65,500	85,000	83,500	155,000

TABLE 23.7 Values of Gt for use in calculating deflection of wood structural panel shear walls and diaphragms. *(continued)*

	Thickness (in.)	OTHER			STRUCTRAL I		
		A-A, A-C	Marine	All Other Grades	A-A, A-C	Marine	All Other Grades
Sanded Plywood	1/4	24,000	31,000	24,000	31,000	31,000	31,000
	11/32	25,500	33,000	25,500	33,000	33,000	33,000
	3/8	26,000	34,000	26,000	34,000	34,000	34,000
	15/32	38,000	49,500	38,000	49,500	49,500	49,500
	1/2	38,500	50,000	38,500	50,000	50,000	50,000
	19/32	49,000	63,500	49,000	63,500	63,500	63,500
	5/8	49,500	64,500	49,500	64,500	64,500	64,500
	23/32	50,500	65,500	50,500	65,500	65,500	65,500
	3/4	51,000	66,500	51,000	66,500	66,500	66,500
	7/8	52,500	68,500	52,500	68,500	68,500	68,500
	1	73,500	95,500	73,500	95,500	95,500	95,500
	1 1/8	75,000	97,500	75,000	97,500	97,500	97,500

For SI: 1 inch = 25.4 mm, 1 pound/inch = 0.1751 N/mm.

a. Applies to plywood with five or more layers; for five-ply/three-layer plywood, use values for four ply.

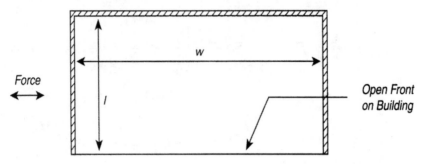

FIGURE 23.1 Diaphragm length and width for plan view of open-front building.

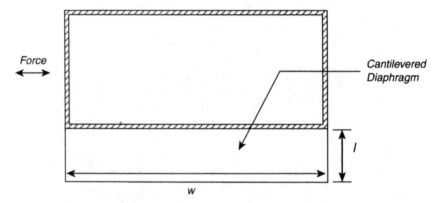

FIGURE 23.2 Diaphragm length and width for plan view of cantilevered diaphragm.

CONVENTIONAL LIGHT-FRAME CONSTRUCTION

As the title suggests, this section of the code covers conventional light-frame construction. Other methods are allowed to be used, but only if you submit a satisfactory design that is in compliance. If your building or structure is a one-, two-, or multiple-family home, such as a townhouses, is no more than three stories above grade plane, and has a separate means of exit, it must comply with the International Residential Code.

Make note that there is a change regarding the design of engineered elements or systems. They are allowed but are subject to certain limits if they exceed the limitations of conventional construction. When this happens, these elements and the supporting load path must be designed in accordance with accepted engineering practice and the provisions of the code. These limitations also apply to other structural elements or systems that are not described in the code, which must also be designed in accordance with accepted engineering practice.

TABLE 23.8 Allowable loads for lumber decking.

Pattern	ALLOWABLE AREA LOAD[a,b]	
	Flexure	**Deflection**
Simple span	$\sigma_b = \dfrac{8F_b'}{l^2}\dfrac{d^2}{6}$	$\sigma_\Delta = \dfrac{384\Delta E'}{5l^4}\dfrac{d^3}{12}$
Two-span continuous	$\sigma_b = \dfrac{8F_b'}{l^2}\dfrac{d^2}{6}$	$\sigma_\Delta = \dfrac{185\Delta E'}{l^4}\dfrac{d^3}{12}$
Combination simple- and two-span continuous	$\sigma_b = \dfrac{8F_b'}{l^2}\dfrac{d^2}{6}$	$\sigma_\Delta = \dfrac{131\Delta E'}{l^4}\dfrac{d^3}{12}$
Cantilevered pieces intermixed	$\sigma_b = \dfrac{20F_b'}{3l^2}\dfrac{d^2}{6}$	$\sigma_\Delta = \dfrac{105\Delta E'}{l^4}\dfrac{d^3}{12}$
Controlled random layup		
Mechanically laminated decking	$\sigma_b = \dfrac{20F_b'}{3l^2}\dfrac{d^2}{6}$	$\sigma_\Delta = \dfrac{100\Delta E'}{l^4}\dfrac{d^3}{12}$
2-inch decking	$\sigma_b = \dfrac{20F_b'}{3l^2}\dfrac{d^2}{6}$	$\sigma_\Delta = \dfrac{100\Delta E'}{l^4}\dfrac{d^3}{12}$
3-inch and 4-inch decking	$\sigma_b = \dfrac{20F_b'}{3l^2}\dfrac{d^2}{6}$	$\sigma_\Delta = \dfrac{116\Delta E'}{l^4}\dfrac{d^3}{12}$

For SI: 1 inch = 25.4 mm

a. σ_b = Allowable total uniform load limited by bending.
σ_Δ = Allowable total uniform load limited by deflection.
b. d = Actual decking thickness.
l = Span of decking
F_b' = Allowable bending stress adjusted by applicable factors.
E' = Modulus of elasticity adjusted by applicable factors.

TABLE 23.9 Diagonally sheathed lumber diaphragm nailing schedule.

SHEATHING NOMINAL DIMENSION	NAILING TO INTERMEDIATE AND END-BEARING STUDS		NAILING AT THE SHEAR PANEL BOUNDARIES	
	Type, size and number of nails per board			
	Common nails	Box nails	Common nails	Box nails
1 × 6	2-8d	3-8d	3-8d	5-8d
1 × 8	3-8d	4-8d	4-8d	6-8d
2 × 6	2-16d	3-16d	3-16d	5-16d
2 × 8	3-16d	4-16d	4-16d	6-16d

Floor Joists

The spans for floor joists must comply with the table figures listed in Chapter 23 of the *International Building Code 2006.*

For other spans and wood species that you cannot locate in these tables, please refer to AF&PA Span Tables for Joists and Rafters. The ends of each joist, other than where supported on a 1-inch-by-4-inch ribbon strip and nailed, cannot have less than 1 1/2 inches or more than 3 inches bearing on wood or metal. Make sure that all joists are supported laterally at the ends and at each support by solid blocking except where the ends of the joists are nailed to the header.

If your joist framing is on opposite sides of a beam or girder, make sure that it is lapped at least 3 inches; if not lapped, then tie the opposite ends together; the manner in which the ends are tied must be approved. A joist must be cut away to give way for floor openings. Specifications usually require that headers be double and framed between the full-length joists, also known as trimmers, on either side of the floor opening.

The ends of header joints longer than 6 feet have to be supported by framing anchors or joist hangers unless the header joints are bearing on a beam or wall.

Braced Wall Lines and Panels

A braced-wall line is a series of braced-wall panels in a single story. All braced-wall lines must meet the requirements of the IBC. Such requirements include location, type, and amount of bracing and are included in Figure 23.3 and specified in Table 23.10.

There are several methods that you can use to construct braced-wall panels:

• Wood boards of 5/8 inch minimum thickness applied diagonally on studs with spacing of not more than 24 inches
• Fiberboard sheathing panels not less than 1/2 inch applied vertically or horizontally on studs with spacing of not more than 16 inches
• Portland-cement plaster on studs spaced 16 inches apart

Don't forget that ceiling joists must have a minimum allowable span. Tables in Chapter 23 of the *International Building Code 2006* contain such spans for specific species and grades of lumber.

Purlins

Purlins that are installed for roof-load support can be used to reduce the span of rafters within the allowable limits. They must be supported by struts attached to bearing walls, and under no circumstance can the purlin be smaller than the rafter that it is supporting. This horizontal structural member must adhere to the allowable limits. For example, a purlin that is 2 inches by 4 inches has a maximum length of 4 feet; a purlin that is 2 inches by 6 inches has a maximum length of 6 feet.

Engineered-Wood Products

A number of changes in the 2006 International Building Code relate to engineered-wood products, which can be composite lumber, prefabricated I-joists, and structural glue-laminated timber. They cannot be notched or drilled unless the manufacturer has given permission to do so or if the effects of doing so are specifically stated in the design by a registered design professional.

Seismic Requirements

There are additional requirements for conventional construction in Seismic Design Categories B and C. In Seismic Design Category C structures of conven-

!Definitionalert

Floor Joists: The main sub-floor framing members that support the floor span.

SEISMIC DESIGN CATEGORY	MAXIMUM WALL SPACING (feet)	REQUIRED BRACING LENGTH, b
A, B and C	35'-0"	Table 2308.9.3(1) and Section 2308.9.3
D and E	25'-0"	Table 2308.12.4

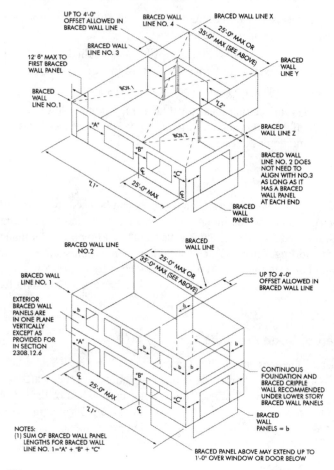

For SI: 1 foot = 304.8 mm.

FIGURE 23.3 Basic components of the lateral bracing system.

TABLE 23.10 Braced wall panels.[a]

SEISMIC DESIGN CATEGORY	CONDITION	CONSTRUCTION METHODS[b,c]								BRACED PANEL LOCATION AND LENGTH[d]
		1	2	3	4	5	6	7	8	
A and B	One story, top of two or three story	X	X	X	X	X	X	X	X	Located in accordance with Section 2308.9.3 and not more than 25 feet on center.
	First story of two story or second story of three story	X	X	X	X	X	X	X	X	
	First story of three story	—	X	X	X	X[e]	X	X	X	
C	One story or top of two story	—	X	X	X	X	X	X	X	Located in accordance with Section 2308.9.3 and not more than 25 feet on center.
	First story of two story	—	X	X	X	X[e]	X	X	X	Located in accordance with Section 2308.9.3 and not more than 25 feet on center, but total length shall not be less than 25% of building length.[f]

For SI: 1 inch = 25.4 mm, 1inch = 304.8 mm.

a. This table specifies minimum requirements for braced panels that form interior or exterior braced wall lines.

b. See Section 2308.9.3 for full description.

c. See Sections 2308.9.3.1 and 2308.9.3.2 for alternative braced panel requirements.

d. Building length is the dimension parallel to the braced wall length.

e. Gypsum wallboard applied to framing supports that are spaced at 16 inches on center.

f. The required lengths shall be doubled for gypsum board applied to only one face of a braced wall panel.

tional light-frame construction cannot be more than two stories in height. Concrete or masonry walls cannot extend above the basement, although there are exceptions:

• You may use masonry veneer in the first two stories above grade plane or, in Seismic Design Category B, the first three stories where the lowest story is of concrete or masonry but only if panel wall bracing is used and is 1 1/2 times the required length (see Table 23.10).

• You may use masonry veneer in the first story above grade plane or the first two stories above grade plane where the lowest story has concrete or masonry walls in Seismic Design Category B or C.

• You may use masonry veneer in the first two stories above grade plane in Seismic Design Categories B and C if the type of brace used in Section 2308.9.3 is method 3 and the bracing of the top story is located at each end and at least every 25 feet. You must also provide hold-down connectors at the ends of braced walls from the second floor to the first floor assembly with a design load of 2,000 pounds.

• You may not use cripple walls.

In the remainder of this chapter I will be discussing irregular structures. You may not use conventional light-frame construction in irregular parts of structures in Seismic Design Category D or E. As you are aware, Chapter 16 covers the forces in that irregular portions of structures are designed to resist.

There are six conditions in which a portion of a building or structure will be deemed irregular. If your building or structure contains one or more of these conditions, you have an irregular structure. The list of such conditions is explained below:

• Exterior braced-wall panels are required to be in one plane vertically from the foundation to the uppermost story. If this is not true of your structure, it will be considered irregular. (See Figure 23.4.) As with many sections of this book, this requirement has exceptions. Floors with cantilevers that are not more than four times the depth of the floor joists are permitted provided that the following is true:

  - Floor joists are 2 inches by 10 inches or larger
  - The back span to the cantilever has a ratio of 2:1

!Definitionalert

Purlins: Framing members that support a roof-panel assembly.

!Codealert

Stress grading of structural log members of nonrectangular shape, as typically used in log buildings, must be used in accordance with ASTM D 3957, and round timber poles and piles must comply with ASTM D 3200 and ASTM D 25.

- Floor joists at the ends of the braced walls are doubled
- A continuous rim joint is connected to the ends of cantilevered joists
- The ends of cantilevered joists do not carry gravity loads from more than a single story

• Any section of a floor or roof that is not laterally supported by braced-wall lines on all edges will be considered irregular except that portions of roofs or floors that do not support braced-wall panels are allowed to extend up to 6 feet. (See Figure 23.5 and 23.6.)

• If the end of a required braced-wall panel is more than 1 foot over an opening in the wall below, your structure is irregular. Braced-wall panels are allowed to extend more than 8 feet in width if the header is 4 inches by 12 inches or larger. (See Figure 23.7.)

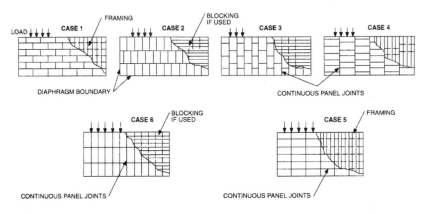

FIGURE 23.4 Allowable shear examples as they may appear in your code book.

- If portions of a floor level are vertically offset so the framing members on either side cannot be lapped or tied, your structure is irregular. Framing supported directly by foundations does not need to be lapped or tied. (See Figure 23.7.)

- If braced-wall lines are not perpendicular to each other, you have an irregular structure.

- Openings in floor and roof diaphragms that have a maximum dimension more than 50 percent of the distance between lines of bracing will put your building in the irregular-structure category. (See Figure 23.8.).

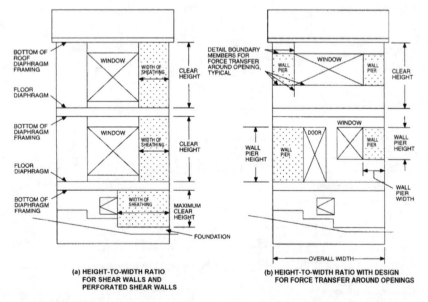

FIGURE 23.5 General definition of shear wall height, width, and height-to-width ratio.

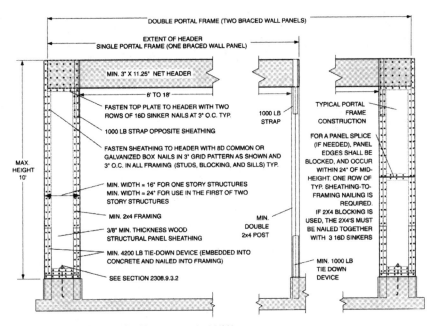

For SI: 1 foot = 304.8 mm; 1 inch = 25.4 mm; 1 pound = 4.448 N.

FIGURE 23.6 Alternate braced wall panel adjacent to a door or window opening.

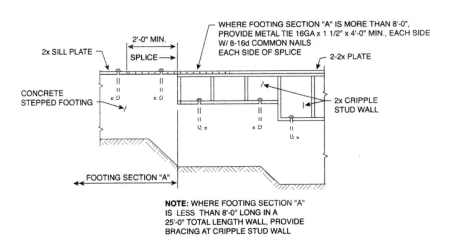

For SI: 1 inch = 25.4 mm, 1 foot = 304.8 mm.

FIGURE 23.7 Stepped footing connection details.

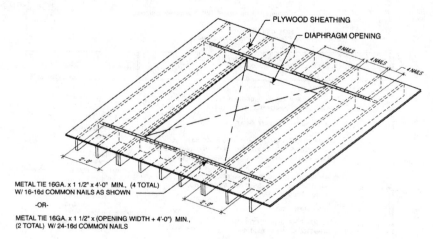

METAL TIE 16GA. x 1 1/2" x 4'-0" MIN., (4 TOTAL)
W/ 16-16d COMMON NAILS AS SHOWN

-OR-

METAL TIE 16GA. x 1 1/2" x (OPENING WIDTH + 4'-0") MIN.,
(2 TOTAL) W/ 24-16d COMMON NAILS

For SI: 1 inch = 25.4 mm, 1 foot = 304.8 mm.

FIGURE 23.8   Openings in horizontal diaphragms.

CHAPTER 24
GLASS AND GLAZING

This chapter describes provisions that must be met when working with glass and glazing. It includes materials, design, construction, and the quality of glass. You will also find provisions for glazing, the installation of glass in handrails and guards, and many more important details that you must be aware of when working on such projects.

GENERAL REQUIREMENTS FOR GLASS

General requirements for glass include identification, supports, framing, interior glazed areas, and louvered windows or jalousies. Each of these requirements is discussed in this section.

Like most materials that I've discussed in this book, glass, too, must include the manufacturer's mark that appropriately identifies the type and thickness of the glass or glazing material. The only time that the identification can be absent is if the glass is approved by the code official and an affidavit is provided by the glazing contractor certifying that the glass used is in accordance with approved construction documents. With the exception of tempered spandrel glass, each and every pane of tempered glass must be permanently marked by the manufacturer. There are several ways that this is done.

The identification mark can be acid-etched, sandblasted, or embossed. The important thing to remember regarding these marks is that once they are applied, they cannot be removed without breaking the glass. The reason why tempered spandrel glass doesn't need a permanent mark is because this type of glass comes with a removable paper mark.

Another requirement for glass is that it needs supports when it is not fully supported on more than one side of a pane or has unusual load conditions. In either of these cases a registered design professional must prepare documents such as detailed shop drawings and analysis or test data to assure safe performance of such glass.

Framing is also a requirement for glass and must be designed so that each individual pane is fully supported and that the deflection of the edge of the glass perpendicular to the pane does not exceed 3/4 inch of the edge. Section 1605 of the 2006 International Building Code has provisions for positive and negative loads that glass may be subject to. It would be wise to refer to this when working with glass.

Be sure that you know the provisions for interior glazed areas. If interior glazing is installed next to a walking surface, the deflection of two adjacent unsupported edges cannot be more than the thickness of the panels if a force of 50 pounds (per linear foot) is applied horizontally to one panel above the walking space.

The last general requirement for glass is in regard to louvered windows or jalousies. There are three types of glass used in both windows and jalousies: float, wired, and patterned. These glasses cannot be any thinner than nominal 3/16 inch and no longer than 48 inches. All glass edges in louvered windows and jalousies must be smooth. You may not use glass with wire exposed on longitudinal edges for louvered windows or jalousies. If you choose to use other types of glass for either louvered windows or jalousies, you must be sure to submit any designs to the building official for prior approval.

Glass Loads

Any glass that is sloped 15 degrees or less from vertical in windows, doors, curtains, or other exterior applications must be designed to resist wind loads. See Section 1609 for these components. You must also refer to ASCE 7, Section 13.5.9, for the seismic requirements for glass. ASTM E 1300 contains the resistance of glass under uniform loads. I suggest that you refer to these standards before installing glass.

The following list contains glass designs for which equations must be used before installing:

!Definitionalert

Jalousies: A series of horizontal glass slats held at each end by movable metal frames attached to each other by louvers.

doesn't need to be installed in handrails or guards in parking garages except for pedestrian areas that not are exposed to impact from vehicles.

ATHLETIC FACILITIES

Glazing used in gymnasiums and in racquetball, squash, and basketball courts as part of athletic facilities must comply with this section. This includes any glazing that is part of or a whole door. In areas such as racquetball and squash courts, test methods and loads for individual glazed areas must conform to CPSC 16 CFR, Part 1201 (see Chapter 35), for loads with impacts applied at a height of 59 inches above the playing surface to glass wall installation. This must be applied to the actual glass and to a simulated glass wall used for practice.

Glass walls used in a racquetball or squash courts must not break following the impact from testing, and the deflection of such walls cannot be greater than 1 1/2 inches at the point of impact for a drop height of 48 inches.

Glass doors must remain intact after a test impact at the prescribed height in the center of the door, and the relative deflection between the edge of the door and the nearby wall cannot be more than the thickness of the wall plus 1/2 inch for a drop height of 48 inches.

All glazing in gymnasiums, basketball courts, or other similar facilities that are subject to human impact loads must comply with Category II of CPSC 16 CFR 1201; again, this can be found in Chapter 35.

The last comment I need to make before I end this chapter is in regard to elevator enclosures. All glass in elevator enclosures must be laminated glass and conform to ANSI Z97.1 or 16 CFR Part 1201, and all markings as specified in the applicable standard must be on each separate piece of glass and visible even after installation.

CHAPTER 26
PLASTIC

Plastic components are a part of construction that is becoming more prevalent. The types of plastics that are included in this section are listed below:

- Foam plastic
- Foam-plastic insulation
- Plastic veneer
- Interior plastic finish and trim
- Light-transmitting plastics

FOAM-PLASTIC INSULATION

When your package of foam-plastic insulation arrives at the job site, make sure that the label of an approved agency is attached to the package or container and also that the manufacturer's name, the product listing, the product identification, and any information that will be used to identify the end use is present. Both foam-plastic insulation and foam-plastic cores of manufactured assemblies will have a flame-spread index of not more than 75. A smoke-developed index of not more than 450 must also be a characteristic of foam insulation, and both must be in accordance with ASTM E 84. All loose-fill-type foam plastic has to be tested as board stock for the flame-spread index and smoke-developed index. A list of exceptions is as follows:

- Section 2604.2 contains the smoke-developed index for interior trim.

- In buildings that are used as cold storage for items such as ice plants, food plants, or food-processing rooms where insulation normally has a thickness of 4 inches, up to 10 inches will be allowed if the building is equipped throughout with an automatic fire sprinkler.

- Roof-covering assemblies such as Class A, B, or C that contain foam-plastic insulation are allowed provided that the insulation passes FM 4450 or UL 1256.
- Foam-plastic insulation that is greater than 4 inches in thickness must have a maximum flame index of 450 when tested at a minimum thickness of 4 inches.
- Flame-spread and smoke-developed indexes for foam-developed-plastic interior signs in covered mall buildings are allowed provided that the signs comply with section 402.15.

With the exception of masonry or concrete construction and cooler and freezer walls, foam plastic must be separated from the inside of the building by a thermal barrier of 1/2-inch gypsum wallboard to limit the average temperature rise of the unexposed surface to no more than 250 degrees F after 15 minutes of fire exposure, complying with the standard time-temperature curve of ASTM E 119. As I have said previously, a thermal barrier is not needed in masonry or concrete construction. This is true if the concrete wall, floor, or roof system is covered on each face by a minimum of 1-inch-thick masonry or concrete. A thermal barrier is also not required for cooler and freezer walls that are installed with foam plastic with a thickness of 10 inches that must:

- Have a flame-spread index of 25 or less with a smoke-developed index of not more than 450
- Have flash ignition and self-ignition temperatures of not less than 600 to 800 degrees F
- Have a covering of no less than 0.032-inch corrosion-resistant steel with a base-metal thickness less than 0.0160 inch at any point
- Be protected by an automatic sprinkler system
- Be part of a building that is sprinklered

One-story buildings that use a foam plastic with a flame-spread index of less than 25 do not have to have a thermal barrier in or on outside walls in a thickness not more than 4 inches. This is only true where the foam is covered by no less than a 0.032-inch-thick aluminum steel with a base thickness of 0.0160 inch. The building must be fully equipped throughout with an automatic sprinkler system. Foamplastic insulation that is mounted in accordance with both the code and the manufacturer's instructions must be separated from the inside of the building with a wood-panel sheathing that is no less than 0.47 inch thick and has been bonded with exterior glue.

The following list contains the provisions for doors that are not required to have a fire-protection rating:

- Pivoted or side-hinged doors having a flame-spread index of 75 or less are allowed as a core material.
- Foam-filled exterior entrance doors in buildings of Group R-2 or R-3 must be faced with wood or another approved material.

• Garage doors with foam plastic used as core material must have a metal door facing with a thickness of 0.032 inch.

There is an exception for garage doors using foam-plastic insulation. Compliance with section 2603.3 for detached and attached garages of one- and two-family dwellings does not require with a thermal barrier.

Whenever foam plastic is used as interior trim and compliance with section 2604 is met, a thermal barrier is not needed. This is also true for interior signs that comply with section 402.15. However, be aware that foam-plastic signs that are not attached to the interior of a building must comply with Chapter 8 of the International Fire Code.

In most cases, foam plastic must be installed with a protective thermal barrier; but as you can see from the above list, there are instances in which a thermal barrier is not needed. Nonetheless, you must make sure that you are in compliance with the appropriate sections that allow these exceptions. If you ever have any questions regarding the installation of either foam plastic or thermal barriers, check with the building official or your local authorities beforehand.

When you apply foam-plastic spray to a sill plate and header of Type V construction, there are three conditions that must be met. Plastic-foam spray has to have a maximum thickness of 3 1/4 inches, and foam plastic has to have a density range of 1.5 to 2.0 pcf and a flame-spread index of 25 or less along with a smoke-developed index of 450 or less when tested with ASTM E 84.

Foam-plastic insulation and exterior coatings and facings must all be tested separately for the thickness that is intended for use and for flame-spread and smoke-developed indexes. ASTM E 84 has determined that the flame-spread index for all three cannot be more than 25 and the smoke-developed index more than 450. But prefabricated or factory-manufactured panels can be tested as an assembly as long as the foam-plastic core is not exposed during construction and if panels have a minimum thickness of 0.020-inch aluminum facing and a total thickness of 0.25 inch or less.

!**Definition**alert

Foam-Plastic Insulation: A plastic that has the ability to expand through the use of a foaming agent that produces a reduced-density material that contains voids consisting of open or closed cells that are distributed throughout for thermal insulating purposes and with a density less than 20 pounds per cubic foot.

There has been a change in the code in regard to using foam plastics as protection against termites. Figure 2603.8 of the *International Building Code 2006* shows a map defining areas of termite infestation. The keyshows that the northern area has a probability of none to slight, while the southern area has a heavy probability of termites.

It is now known that termites have a tendency to burrow through expanded polystyrene and other foam plastics to get to wood, therefore exterior foam plastics may not be installed below grade. But there are still a few exceptions as found below:

• In buildings where walls, floors, and ceilings are made of wood that has been treated with a preservative or noncombustible materials

• An approved method is used to protect foam plastic and structures from termite damage

• On the inside of basement walls

Foam plastics are not required to comply with sections 2603.4 through 2603.7 if approval has been met through testing. This testing is not limited to NFPA 286, FM 4880, or UL 140 or 1715, and any testing must be in relation to the use of the foam plastic. When allowed for use as interior finish on the basis of special testing, you must also refer to Chapter 8 for compliance with the flame-spread index. Testing will include seams, joints, and other details typically used for this type of installation. Testing must also include the manner intended for use.

INTERIOR FINISH AND TRIM

Plastic materials must comply with this section and Chapter 8 when installed as interior finish and trim. Also refer to the end of the above paragraph as a reminder that special approval is required. When and if you've received special approval to use plastic materials for trim, you must also comply with the following:

• Interior trim must have a minimum density of 20 pcf.

• Interior trim must have a maximum thickness of 0.5 inch and a maximum width of 8 inches.

• There will be an area limitation of no more than 10 percent of the aggregate wall and ceiling area of any room or space.

• When tested in accordance with ASTM E 84, the flame-spread index may not be more than 75.

• No limitation is made on the smoke-developed index.

PLASTIC VENEER

Plastic veneer used inside a building must be in compliance with interior-finish and trim requirements and Chapter 8. You are allowed to install exterior plastic veneer on the exterior walls of buildings and are not restricted to the type of construction; however, there are guidelines that must be met:

- Compliance with section 2606.4 is a must.
- Do not attach plastic veneer to any outside wall with a height of more than 50 feet above grade.
- There is a 300-square-foot minimum for sections of plastic veneer with a separation minimum of 4 feet.
- Note that the area and separation requirements and the smoke-density limitation do not apply to Type V-B construction, provided that there is not a fire-resistance-rating requirement for walls.

LIGHT-TRANSMITTING PLASTICS

Before you use any light-transmitting materials, you must first submit for approval sufficient data to demonstrate the use of such materials. It is up to the building official to allow the use of light-transmitting materials in your building or structure. Light-transmitting plastics include thermoplastic, thermosetting, or reinforced thermosetting plastic materials, and all of these must have a self-ignition temperature of 650 degrees F or higher when tested in accordance with ASTM D 1929.

The smoke-developed index cannot be greater than 450 when tested in accordance with ASTM E 84 or greater than 75 when tested in accordance with ASTM D 2843. All light-transmitting materials must conform to either Class CC1, in which plastic materials have a 1 inch or less burning extent when tested at a thickness of 0.060 inch, or to Class CC2, in which plastic materials have a burning rate of 2.5 inches per minute when tested at a thickness of 0.060 inch. Both classes must be tested in accordance with ASTM D 635.

As indicated in regard to the durability and loads of Chapter 16, light-transmitting plastic materials must comply with these provisions. The building official will use any data that establishes stresses, maximum weight spans, and other important information in making any decisions about the use of light-transmitting plastics. Submission of this information and data is mandatory on your part; not doing so means not obtaining approval from the building official.

There are certain occupancies and locations in which a light-diffusing system cannot be installed without an automatic sprinkler system:

- Group A occupancies with an occupant load of 1,000 or more
- Theaters that have a stage and proscenium opening and an occupant load of more than 700
- Group I-2 and I-3 occupancies
- Stairways and passageways intended for exits

Hangers of at least No. 12 steel must be used as support for light-transmitting diffusers. They must be hung directly or indirectly from ceilings or roofs, and the hangers have to be of a noncombustible material. Be sure that you are in compliance with Chapter 8 unless the plastic diffusers have the ability to fall from their mountings before ignition takes place at a temperature of 200 degrees F; this temperature is below the ignition temperature of panels. Please note that panels cannot be more than 10 feet in length or 30 square feet in area.

You are allowed to install light-transmitting plastics as glazing in shower stalls, doors, and bathtub enclosures as long as you are in compliance with Chapter 24.

Light-Transmitting Plastic Wall Panels

You cannot use light-transmitting plastic as wall panels in exterior walls in occupancies of the following groups:

- A-1
- A-2
- H
- I-2
- I-3

In other groups panels are allowed but only if the walls are not required to have a fire-resistance rating and installation conforms to this section. For instance, exterior wall panels cannot alter the type of construction. 26.1 specifies the area limitation and separation requirements for light-transmitting plastic wall panels.

Combinations of light-transmitting plastic glazing and light-transmitting plastic wall panels are subject to height, area, and percentage limitations along with any separation requirements.

Light-Transmitting Plastic Glazing

If you remember correctly, there are provisions in Chapter 7 regarding the protection for openings in exterior walls of Type V-B construction. However, areas not required to comply with that chapter are permitted to have light-transmitting plastic as long as it in compliance with its proper section of this chapter and also in accordance with the following:

- The total area of the plastic glazing cannot be more than 25 percent of any area of any wall face of the story in which it is installed, except for those areas that have an automatic sprinkler system; only then can the total area be increased to a maximum of 50 percent.

- Installation of an approved flame barrier between glazed units in adjacent stories is allowed and must extend 30 inches beyond the outside wall or vertical panel no less than 4 feet. An exception to this rule is a building equipped throughout with an automatic sprinkler system.

TABLE 26.1 Area limitation and separation requirements for light-transmitting plastic wall panels.[a]

FIRE SEPARATION DISTANCE (feet)	CLASS OF PLASTIC	MAXIMUM PERCENTAGE AREA OF EXTERIOR WALL IN PLASTIC WALL PANELS	MAXIMUM SINGLE AREA PLASTIC WALL PANELS (square feet)	MINIMUM SEPARATION OF PLASTIC WALL PANELS (feet)	
				Vertical	Horizontal
Less than 6	—	Not permitted	Not permitted	—	—
6 or more but less than 11	CC1	10	50	8	4
	CC2	Not permitted	Not permitted	—	—
11 or more but less than or equal to 30	CC1	25	90	6	4
	CC2	15	70	8	4
Over 30	CC1	50	Not limited	3[b]	0
	CC2	50	100	6[b]	3

For SI: 1 foot = 304.8 mm, 1 square foot = 0.0929 m^2.
a. For combinations of plastic glazing and plastic wall panel areas permitted , see Section 2607.6.
b. For reductions in vertical separation allowed, see Section 2607.4.

• Do not install light-transmitting plastics more than 75 feet above grade plane. Again, an exception to this rule is a building equipped throughout with an automatic sprinkler system.

Light-Transmitting Plastic Roof Panels

Light-transmitting plastic roof panels panels cannot be installed in Groups H, I-2, and I-3. You may install them in other groups as long as they comply with any one of the following conditions:

• The building has an automatic sprinkler system throughout.
• The roof is not required to comply with the fire-resistance rating as prescribed by Table 6.1.
• Roof panels meet the requirements for roof coverings in Chapter 15.

The total area of roof panels is limited by a percentage of the floor area. Table 26.2 contains the limitations; exceptions include the following:

• Area limitations are increased by 100 percent if the building has an automatic sprinkler system.
• Low-hazard-occupancy buildings are exempt from Table 26.2 limitations provided that the buildings are not larger than 5,000 square feet and have a minimum fire-separation distance of 10 feet.
• Greenhouses without public access are exempt from area limitations but only if they have a minimum fire-separation distance of 4 feet.
• Roof coverings of terraces and patios in Group R-3 occupancies are permitted to use light-transmitting plastics.

Light-Transmitting Plastic Skylight Glazing

This section covers the provisions for skylight assemblies using light-transmitting plastic. If installing unit skylights that have been glazed with light-transmitting plastic, be sure to refer to Chapter 24 for compliance requirements. This provision

TABLE 26.2 Area limitations for light-transmitting plastic roof panels.

CLASS OF PLASTIC	MAXIMUM AREA OF INDIVIDUAL ROOF PANELS (square feet)	MAXIMUM AGGREGATE AREA OF ROOF PANELS (percent of floor area)
CC1	300	30
CC2	100	25

For SI: 1 square foot = 0.0929 m².

does not include skylights that contain light-transmitting plastics that already conform to Chapter 15. Light-transmitted plastics are to be set upon curbs that have been constructed above the plane of the roof by at least 4 inches. Make sure that all edges are protected by metal or another material that has been approved for protection.

Any material that you use must be ignition-resistant to exposure to flames. Curbs are not required for skylights that are used on roofs that have a minimum slope of 3 units vertical in 12 units horizontal. This pertains to occupancies in Group R-3 and buildings that have a noncombustible roof covering. And metal or noncombustible edge materials are not required where nonclassified roof coverings are allowed.

Skylights may not exceed a maximum of 100 square feet within the curb area, except that this limitation does not apply to buildings that are equipped with an automatic sprinkler system. Make sure that you have placed all your skylights at a distance of no less than 4 feet measured in a horizontal plane.

Buildings with an automatic sprinkler system and Group R-3 occupancies that have more than one skylight that doesn't exceed the area limitations are not included in the separation provisions.

Light-Transmitting-Plastic Interior Signs

Light-transmitting interior signs have to follow the provisions of both Chapters 2 and 4 of the IBC and cannot have an area more than 20 percent of the wall area, with a maximum area of 24 square feet. The edges and backs of all plastic interior signs must be fully encased in metal.

CHAPTER 29
PLUMBING SYSTEMS

The plumbing aspects of the 2006 International Building Code, which include the erection, installation, alteration, and maintenance of plumbing equipment and systems, must be designed and constructed in accordance with the International Plumbing Code. Private sewage disposal must comply with the International Private Sewage Disposal Code. Please consult these referenced standards for all plumbing questions.

CHAPTER 31
SPECIAL CONSTRUCTION

What comes to mind when someone mentions special construction? Isn't all construction special? Actually, special construction involves membrane structures, temporary structures, walkways, towers, and antennas. There are probably many more types of special construction that you can come up with. With all types of special construction you can bet there is a provision that must be followed. That's what this chapter is about.

MEMBRANE STRUCTURES

Membrane-covered cable and membrane-covered frame structures are known as membrane structures. These are air-supported and air-inflated, and they are usually erected for a length of 180 days or longer. However, if they are up for less than 180 days, be sure that you've consulted the International Fire Code for other provisions that must be complied with. Certain membrane structures only need to meet the requirements for membrane and interior-liner material and engineering design. These are membrane structures that are not used for human occupancies:

• Water-storage facilities

• Water clarifiers

• Water-treatment plants

• Sewage-treatment plants

• Greenhouses

Type II-B construction, such as membrane and interior-liner material, describes noncombustible membrane structures. Type IV construction is concerned with heavy timber-frame-supported structures that are covered by an approved membrane, while Type V construction defines all other membrane structures. The exception is plastic than is less than 30 feet above any floor or less than 20 mil and used in greenhouses and for aquaculture pond covers, which is not required to meet the fire-propagation performance requirements of NFPA 701.

Liners must be either noncombustible as set forth in Chapter 7 or meet NFPA 701. If you go back to Table 4.4, you will find the maximum floor areas and maximum height requirements for membrane structures. Chapter 16 contains provisions for engineering design regarding dead loads, loads due to tension, and live loads including wind, snow, flood, and seismic loads.

There at least one blower in an inflation system, and the system must have provisions for automatic control to maintain inflation pressures. The design of the system must also prevent overpressurization. In buildings that are more than 1,500 square feet make sure you've installed an auxiliary inflation system for maintenance of the structure in case the primary system fails. The auxiliary system must be automatic so that there is no loss of internal pressure when the primary blower doesn't work. The blower equipment must meet all five of the following requirements:

• A continuous-rated motor must power all blowers.

• Protective devices such as inlet screens or belt guards are required to provide protection; the building official may require others.

• All blowers must be housed in a weather-protected structure.

• All blowers must be equipped with a backdraft damper to minimize air loss when the blower isn't working.

• Blower inlets must be placed to provide protection from contaminated air. All locations must be approved by your building official.

It only makes sense to require a backup power-generating system when you need an auxiliary inflation system. Equip the system with a means of automatic starting power upon failure of normal service within 60 seconds for a total use of 4 hours.

!Definitionalert

Air-Inflated Structure: A building where the shape of the structure is maintained by air pressurization of cells or tubes to form a barrel vault over the usable area. Occupants of this type of structure do not occupy the pressurized area used to support the structure.

TEMPORARY STRUCTURES

Any building or structure that is erected for less than 180 days is considered temporary and must comply with the International Fire Code. You must contact the building official for a permit if any temporary structure is going to be used for gatherings of more than 10 persons and the structure is over 120 square feet. When you are in need of this type of permit, you have to fill out an application and submit any construction documents. Include the site plan that indicates the location, means of egress, and the occupant load.

PEDESTRIAN WALKWAYS AND TUNNELS

The interior of a building and a walkway must be separated by fire-barrier walls with a fire-resistance rating no less than 2 hours. Fir protection must extend from a point that is 10 feet above the walkway roof surface to a point 10 feet below the walkway and another 10 feet horizontally. The exception to this pertains to the walls separating the walkway from a connected building, which are not required to have a fire-resistance rating when any of the following conditions exist:

• With the exception of open parking garages, connecting buildings and pedestrian walkways must have a fire-separation distance of more than 10 feet, an automatic sprinkler system. and a wired or laminated glass wall that is also protected with a sprinkler system. The glass must be in a gasketed frame and installed to deflect without breaking. There must not be any obstruction installed between the sprinkler heads and the glass.

• A distance of 10 feet must be between the connected building and the walkway, and both sidewalls of the walkway must be open by at least 50 percent.

• Buildings must be on the same lot.

• Exterior walls of connected buildings that are required to have fire-resistance ratings of 2 hours must have walkways equipped with an automatic sprinkler system in accordance with NFPA 13.

A tunneled walkway and the connecting building must have a separation with no less than 2-hour fire-resistant construction.

AWNINGS AND CANOPIES

When you install an awning or a canopy, it must be designed and constructed to withstand wind or other lateral and live loads as required by Chapter 16. Members must be protected to prevent deterioration, and awnings must have

---

!Definitionalert

Retractable Awning: A cover with a frame that retracts against a building to which it is supported.

---

frames of noncombustible material, fire-retardant-treated wood, or 1-hour fire-resistant construction.

Canopy materials have to be constructed of a rigid framework with an approved covering that meets the fire-performance criteria of NFPA 701 or has a flame-spread index no greater 25 when tested in accordance with ASTM E 84.

MARQUEES

A marquee is a permanent roofed structure that is attached to and supported by the building and also projects from the building. A marquee must have a maximum height or thickness (measured vertically) of no more than 3 feet. If the roof or any part is a skylight, refer to Chapter 24 for requirements. Both the roof and the skylight of a marquee have to be sloped to downspouts so that drainage does not end up on the sidewalk. Do not place a marquee so that it interferes with the operation of any outside standpipe or exit discharge from the building. Chapter 16 contains additional information about the design of marquees.

RADIO AND TELEVISION TOWERS

Towers must always be accessible for inspections by the use of step bolts and ladders. When putting up guy wire, make sure that it doesn't cross over any above-ground electric-utility lines or invade private property without written consent of the property owner. A corrosion-resistant noncombustible material must be used in the construction of towers not more than 100 feet high, and construction will be Type IIB.

Towers must be designed to resist wind loads in accordance with TIA/EIA-222. When you erect tower for loads, take ice and snow into consideration. Also make sure that you've grounded the tower.

SWIMMING-POOL ENCLOSURES

Public swimming pools are to be completely enclosed with a fence that is at least 4 feet tall with an opening in the fence that allows the passage of a 4-inch-diameter sphere. The enclosure of the fence must have self-closing and self-latching gates. A residential swimming pool must have a barrier of at least 48 inches above grade with an opening that will allow a 4-inch-diameter sphere. If the barrier surface has decorative cutouts within vertical members, they cannot be more than 1 3/4 inches wide.

If you've installed a chain-link fence around your pool, be aware that the maximum size must be 2 1/4 inches square unless the fence has slats fastened at the top or the bottom that reduce the openings to no more than 1 3/4 inches.

If you're using a dwelling wall as a barrier, you must conform to at least one of the items in the list below:

• A dwelling wall must have an audible alarm that within 7 seconds goes off after the door is opened and continues to sound for at least 30 seconds. This alarm is required to set itself and also must be hooked up to a touch pad for deactivation but for no more than 15 seconds.

• The pool must have a power safety cover that complies with ASTM F 1346.

• You've provided another means of protection such as self-closing and self-latching devices that have been approved by the administrative authority.

To prevent user entrapment, install suction outlets that produce circulation throughout the pool or spa with covers that conforms to ASME A112.19.8M. The drain grate must be 12 inches by 12 inches. (This does not apply to surface skimmers.)

A minimum of two suction outlets must be used for single- or multiple-pump circulation systems with a minimum horizontal or vertical separation of at least 3 feet. The suction outlets have to be piped in such a way as to allow water to be drawn through a vacuum line to the pump. And, last but not least, vacuum- or pressure-cleaner fittings must be placed so that they are accessible and at least 6 inches but not more than 12 inches below the water level.

CHAPTER 32
ENCROACHMENTS INTO THE PUBLIC RIGHT-OF-WAY

Due to the lack of pertinent information, there is not a commentary chapter regarding encroachments into the public right-of-way. This is not to say that it isn't an important area in the 2006 International Building Code; there is just not enough information to comment on.

CHAPTER 33
SAFEGUARDS DURING CONSTRUCTION

In this chapter provisions for safety during construction and the protection of all surrounding public and private properties are covered.

CONSTRUCTION SAFEGUARDS

All exits, existing structural elements, fire protection, and sanitary safeguards must be maintained during all phases of construction such as remodeling, replacement, alterations, or repairs. The only times when these procedures are not required is when the safety devices themselves are being remodeled, altered, or repaired or when the building that is being worked on is unoccupied. There is a likelihood that waste materials will be found on the site. Such materials must be removed and disposed of to prevent any damage to persons, properties, and public right-of-ways.

DEMOLITION

Always keep any construction documents or schedules in a safe place where you can get them at a moment's notice. Why? Because you never know when the building official is going to ask to see them. If these documents cannot produced when asked for, no work can be done until they are located and the building official has approved them.

Before any demolition plans are set in cement, make sure that you've done your homework and assured that protection of pedestrians is in place. If not, you can't begin any work.

Have you thought of an alternate means of exit before demolishing the party wall, balcony, or horizontal exit? You probably should, and this, too, must be approved. Once you've demolished or removed the building and are surrounded by an empty lot, don't forget that you must fill and maintain it to the existing grade or in accordance with the ordinances of the jurisdiction that has authority. Don't forget to make provisions to prevent damage from any water accumulation in the foundation or any adjacent property. And do not try to disconnect any service-utility connections by yourself. You must have utility connections disconnected and capped in accordance with the approved rules and with any requirements of the authority having jurisdiction.

SITE WORK

Areas around construction sites must be cleaned of any stumps or roots that are buried at least 12 inches into the ground; you must also make sure that any excavation or fill is constructed so that the safety of life or property is maintained. Make sure that all wood forms that have been used in the placement of concrete are removed before the building is occupied. Any loose or casual wood must be moved so that it does not have direct contact with the ground.

Slopes for permanent fill cannot be greater than 50 percent. This is also true for cut slopes. If you need to change these limitations, you must present the results of a soil investigation report to the building official.

Be aware of the code regarding surcharge and fill, which cannot be placed next to any building or structure unless you're positive that any additional loads will be supported by the building or structure. Also be sure that footings are pinned to protect against settlement. For more information on footings refer to chapter 18.

SANITARY PROVISIONS

Sanitary provisions during construction, remodeling, and demolition activities are found in the International Plumbing Code.

PROTECTION OF PEDESTRIANS

It is important that all pedestrians are protected during construction, remodeling, and demolition activities, and you must provide signs that direct pedestrians to alternate routes. Unless the jurisdiction authority tells you not to, you must provide a sidewalk for pedestrian travel in front of the construction site. The standard width of a pedestrian sidewalk is no less than 4 feet. If such sidewalk extends out into the street, it needs to be protected by a directional barricade against traffic. Check with Chapter 11 for more information regarding sidewalks and accessibility issues.

There may be times that construction railings have to be erected; make sure that they are at least 42 inches high and that pedestrians are able to navigate around them. Table 33.1 contains the provisions for the protection of pedestrians.

You may decide that barriers are a better match for your construction site than railings. If you use barriers, there are provisions that must be followed. Barriers must be designed to resist loads as are required in chapter 16 unless they are constructed as follows:

• The top and bottom plates of any barrier must be 2 inches by 4 inches.
• The barrier must be made of wood and be a minimum of 3/4 inch thick.
• Wood structural-use panels have to be bonded with an adhesive that is the identical to the bond used for exterior wood structural-use panels.
• Wood structural panels thicker than 5/8 inch cannot span more than 8 feet.

Pedestrian safety is an important issue around construction sites. For construction sites adjacent to sidewalks a covered walkway may be required to protect pedestrians. If you plan on erecting a covered walkway, you must be aware of a number ofrequirements. First, covered walkways must have a minimum clear height of 8 feet measured from the floor to the overhead canopy. Also make sure that you provide adequate lighting at all times. The covered walkway must be de-

TABLE 33.1 Protection of pedestrians.

HEIGHT OF CONSTRUCTION	DISTANCE FROM CONSTRUCTION TO LOT LINE	TYPE OF PROTECTION REQUIRED
8 feet or less	Less than 5 feet	Construction railings
	5 feet or more	None
More than 8 feet	Less than 5 feet	Barrier and covered walkway
	5 feet or more, but not more than one-fourth the height of construction	Barrier and covered walkway
	5 feet or more, but between one-fourth and one-half the height of construction	Barrier
	5 feet or more, but exceeding one-half the height of construction	None

For SI: 1 foot = 304.8 mm.

signed to support a live load no less than 150 psf for the entire structure. There are exceptions to the rules for covered walkways. New, light-frame construction that does not exceed two stories in height is allowed to have roof and supporting structures designed for a live load of 75 psf or the loads imposed on them, whichever is greater. If you do not care for this design, the roof and supporting structure are allowed to be constructed as follows:

• Footings must be continuous 2-inch-by-6-inch members.

• Posts no less than 4 inches by 6 inches must be provided on both sides of the roof, which cannot be spaced no more than 12 feet apart.

• Stringers that are no less than 4 inches by 12 inches have to be placed on the edge upon the posts.

• Joists that are resting on stringers must be at least 2 inches by 8 inches and must be spaced no more than 2 feet apart.

• The deck must be made of planks at least 2 inches thick or wood structural panels with an outside-exposure durability classification at least 23/32 inch thick nailed to the joists.

All pedestrian protection must be repaired, maintained, and kept in good working condition from the moment it is put up to the moment it is taken down. All debris, trash, and other items must be kept clear of any walkways, whether covered or not. When a covered walkway is taken, down the property must be left in the same condition as before it was built. Adjoining properties also must be protected during the construction phase.

Areas that demand protection include footings, foundations, skylights, and roofs. Make sure you control water runoff and erosion to adjoining properties. The person in charge of the excavation must provide adjoining property owners with notice regarding the excavation and that all adjoining buildings must be protected. Said notice must be delivered no less than 10 days before the scheduled starting date of the excavation.

When you bring in construction equipment and materials, be sure to place them so they do not obstruct access to fire hydrants, standpipes, manholes, or other public-works accessories. Nor must any material or equipment be located within 20 feet of street intersections or placed in a way so that traffic signals are hidden from the public view. This is true for the placement of building materials, fences, or other objects that would obstruct the viewing of public accessories, such as fire hydrant, fire-department connections, or manhole covers.

During the construction, alteration, or demolition portable fire extinguishers must be kept on site. They must be found on each stairway on all floors where combustible materials have accumulated and in storage sheds. Make sure that you have additional fire extinguishers where special hazards exist. The IBC and the International Fire Code must be strictly observed in the event of any fire hazard.

• Fire escapes must comply with this section and cannot make up more than 50 percent of the required number of exits, nor can they make up more than 50 percent of the required exit capability.

Fire escapes that are installed on the front and project beyond the building line cannot have a lower landing that is less than 7 feet or more than 12 feet above grade. There must always be a balanced stairway to the street for counterbalance. The ideal fire escape must be designed to support a live load of 100 pounds per square foot and constructed of steel or another approved noncombustible material.

The only time a wood fire escape is allowed is for buildings of Type 5 construction, and it must be at least 2 inches thick. You can have wooden walkways and railings that are over or are supported by combustible roofs in Types 3 and 4 construction, but they too must be at least 2 inches thick.

The stairs for all fire escapes must be at least 22 inches wide with risers that are no more than 8 inches wide and the treads no less than 8 inches wide. Make sure that the landings on all fire escapes are no less than 40 inches wide by 36 inches long.

CHANGE OF OCCUPANCY

You must never make changes in the use or occupancy of any building that would place it in a different division of the same group or in a different occupancy, unless the building complies with the requirements for the new division or group. Any approval for occupancy changes must made by the building official. If the building official gives the okay, a certificate of occupancy will be issued. When this type of occupancy change happens and the structure is reclassified to a higher occupancy category, you must make sure that seismic requirements are in place except for the following:

• If you can show that the level of performance and seismic safety are the same as that of a new structure through analysis that considers the regularity, overstrength, and ductility of the structure, the requirements of the IBC and ASCE 7 do not apply.
• When this change results in a structure being reclassified from Category I or II to Category III and if the structure is located in a seismic map zone where $S_{DS} < 0.33$, compliance with ASCE 7 is not required.

HISTORIC BUILDINGS

It is always nice to see buildings restored to their original condition. This takes time, money, and effort and, of course, a love of history. Some historic buildings are changed from their original occupancy. But did you know that the provisions of the code that relate to the construction, repair, or alterations are not mandatory for historic buildings? They are not, provided, however, that the building official judges that these buildings do not represent a distinct life-safety hazard.

You can continue to make such changes; just check with the building official first. Be aware that historic buildings that are located within flood areas must be in accordance with Chapter 16 if substantial improvements are made, except for the following cases:

- Historic buildings that are determined to be eligible for listing in the National Register of Historic Places

- As determined by the Secretary of the U.S. Department of the Interior, historic buildings contributing to the historical significance of a registered historic district

- Historic buildings designated as historic under a state program and approved by the Department of the Interior

I suggest that you know exactly what is involved before you make any changes in a historic building.

ACCESSIBILITY

This section applies to the accessibility requirements for maintenance, additions, or changes in occupancies to existing buildings, including historic buildings. Not included are Type B dwellings or sleeping units required by section 1107. Buildings, facilities, or elements that are constructed or altered to be accessible must be kept accessible even during occupancy, and such construction of an existing element cannot impose a requirement for greater accessibility than before construction.

When a building is undergoing construction for an occupancy change, the occupancy change must include the following accessible features:

- At least one accessible entrance
- At least one accessible route from an accessible building to primary areas
- If a parking lot is provided, accessible parking must also be provided
- If building zones are provided, at least one accessible passenger zone
- At least one accessible route that connects accessible parking and passenger loading zone to an accessible entrance

If it is ever unfeasible to comply with new construction standards, you must at least conform to the maximum extent that is technically feasible. Changes of occupancy that include alterations or additions must be in compliance with Chapter 3. Where compliance is technically not feasible for the alteration of a building, the alteration must provide access to the maximum extent that is feasible. There are some exceptions to be aware of:

• The altered space is not required to be on an accessible route, unless required by section 3409.7.

• Accessible means of egress as required by Chapter 10 are not required to be provided in existing buildings and facilities.

• Alterations to Type A individually owned dwelling units within a Group R-2 occupancy must meet the provisions for a Type B dwelling unit and must comply with Chapter 11 and ICC/ANSI A117.1.

You will find that there are times when an alteration affects the accessibility to or contains an area of primary function; there must be an accessible route to the primary area of use. This accessible route must include bathroom facilities or drinking fountains that are used by the area of primary function.

The costs of providing such routes are not required to be more than 20 percent of the costs of the alterations. This provision does not apply to alterations that are solely to windows, hardware, electrical outlets, and signs. Nor does it apply to mechanical, electrical, or fire-protection systems.

COMPLIANCE ALTERNATIVES

The last section of this chapter, contains equations and tables that are important to many aspects of the 2006 International Building Code. The first equation, 34.1, is the height formula, used to compute building-height value. Use it along with Chapter 5 to determine the allowable height of the building, which includes allowable increases due to automatic sprinklers.

$$\text{Height value, feet} = \frac{(AH) - (EBH)}{12.5} \times CF$$

(Equation 34.1)

$$\text{Height value, stories} = (AS - EBS) \times CF$$

where:

AH = Allowable height in feet from Table 503.

EBH = Existing building height in feet.

AS = Allowable height in stories from Table 503.

EBS = Existing building height in feet.

CF = 1 if $(AH) - (EBH)$ is positive.

CF = Construction-type factor shown in Table 3410.6.6(2) if $(AH) - (EBH)$ is negative.

TABLE 34.1 Compartmentation values.

		CATEGORIES[a]			
	a	b	c	d	e
OCCUPANCY	Compartment size equal to or greater than (square feet)	Compartment size of 10,000 square feet	Compartment size of 7,500 square feet	Compartment size of 5,000 square feet	Compartment size of 2,500 square feet
A-1, A-3	0	6	10	14	18
A-2	0	4	10	14	18
A-4, B, E, S-2	0	5	10	15	20
F, M, R, S-1	0	4	10	16	22

For SI: 1 square foot = 0.0929 m.^{2}

a. For areas between categories, the compartmentation value shall be obtained by linear interpolation.

The next equation, 34.2, is used to compute the allowable area value, and equation 34.3 is for the area value.

$$AA = \frac{(SP + OP + 100) \times (\text{area, Table 503})}{100}$$ (Equation 34.2)

where:

AA = Allowable area.

SP = Percent increase for sprinklers (Section 506.3).

OP = Percent increase for open perimeter (Section 506.2).

$$\text{Area value } i = \frac{\text{Allowable area}_i}{1,200 \text{ square feet}} \left[1 - \left(\frac{\text{Actual area}_i}{\text{Allowable area}_i} + \ldots + \frac{\text{Actual area}_n}{\text{Allowable area}_n} \right) \right]$$ (Equation 34.3)

where:

i = Value for an individual separated occupancy on a floor.

n = Number of separated occupancies on a floor.

Compartments created by fire barriers or horizontal assemblies must be evaluated by using Table 34.1. Conforming compartments do not include shafts, chases, walls, or columns.

All tenant and dwelling-unit separations (floors and walls) have to be evaluated for fire-resistance ratings by using Table 34.2.

Table 34.2 defines the categories for tenant and dwelling units:

• Category a—No fire partitions; no doors, or doors not self-closing

• Category b—Fire partitions/wall assemblies with less than 1-hour fire-resistance rating

TABLE 34.2 Separation values.

	CATEGORIES				
OCCUPANCY	**a**	**b**	**c**	**d**	**e**
A-1	0	0	0	0	1
A-2	-5	-3	0	1	3
R	-4	-2	0	2	4
A-3, A-4, B, E, F, M, S-1	-4	-3	0	2	4
S-2	-5	-2	0	2	4

• Category c—Fire partitions with 1-hour or greater fire-resistance rating in accordance with Chapter 7

• Floor assemblies with 1-hour but less than 2-hour fire-resistance rating

• Category d—Fire barriers (see category c)

• Category e—Fire barriers and floor assemblies with 2-hour or greater fire-resistance rating

The above list must also be in accordance with Chapter 7. Please refer to Table 34.3 in regard to corridor-wall values.

The smoke-detection capability must be evaluated based on the location and operation of automatic fire detectors in accordance with Chapter 9 and the International Mechanical Code. I've listed the categories for automatic fire detection below (see Table 34.4):

• Category a—None

• Category b—Smoke detectors already in existence in HVAC systems and maintained in accordance with the International Fire Code

• Category c—Smoke detectors in HVAC systems installed in accordance with the International Fire Code

• Category d—Smoke detectors found throughout all floor areas other than sleeping units, tenant spaces, and dwelling units

• Category e—Smoke detectors found throughout the fire area

The means-of-egress capacity and number of exits must also be evaluated. The means of exits must conform to Chapter 10 and to this section.

The last table included in this chapter is the summary sheet on which you will put all of your information regarding evaluation to be submitted to the building official.

TABLE 34.3 Corridor wall values.

OCCUPANCY	CATEGORIES			
	a	b	c[a]	d[a]
A-1	-10	-4	0	2
A-2	-30	-12	0	2
A-3, F, M, R, S-1	-7	-3	0	2
A-4, B, E, S-2	-5	-2	0	5

a. Corridors not providing at least one-half the travel distance for all occupants on a floor shall use Category b.

You have reached the end of this chapter and this book. I do hope that you see the 2006 International Building Code in a different light and understand why you must include the provisions of other codes when construction your building. I wish you good luck with all of your upcoming projects.

TABLE 34.4 Automatic fire detection values.

OCCUPANCY	CATEGORIES				
	a	b	c	d	e
A-1, A-3, F, M, R, S-1	-10	-5	0	2	6
A-2	-25	-5	0	5	9
A-4, B, E, S-2	-4	2	0	4	8

INDEX

A

Accessory occupancies, 5.11
Accessible, 11.2
Accessible entrances, 11.5
Accessible means of egress, 10.6, 10.7
Accessible routes, 11.3
Accessibility requirements, 11.1, 34.4
Additions, 34.1
Administration, 1.1
Agricultural building, 2.1
Aggregate quantities, 4.48
Aircraft-related occupancies, 4.34
Air-inflated structure, 31.2
Aisles, 10.25, 18.21
Alarm systems, 9.14
Alarm-notification appliance, 9.2
Alarm-verification feature, 9.2
Alarms, 9.8
Alarms, emergency, 4.43
Allowable stress design, 23.15
Alternative systems, 9.9
Alteration, 2.1, 34.1
Appeals, 1.6
Application of flammable liquids, 4.53

Approvals, 17.1
Approved, 2.1
Approved source, 2.1
Area, Building, 5.1
Area modifications, 5.5
Areaway, 2.2
Assembly, 3.2, 10.31
Athletic facilities, 24.7
Atriums, 4.10
Attic, 2.2
Audible alarm-notification appliance, 9.2
Automatic sprinkler systems, 4.32, 9.2, 9.6, 33.5
Awning, 2.2, 31.3
Awning, retractable, 31.4

B

Balanced door, 10.8
Basement, 5.1
Bed joint, 21.6
Braced wall lines, 23.21
Braced wall panels, 23.21
Building, 2.2
Building area, 5.1
Building line, 2.2

Building height, 5.2
Building officials, 1.2
Buildings, unlimited-area, 5.7
Business, 3.4

C

Calculated fire resistance, 7.41
Canopy, 2.2, 31.3
Ceiling radiation dampers, 7.37
Ceiling panels, 7.22
Cement plaster, 25.6
Change of occupancy, 34.3
Combustible dust, 4.46
Combustible materials, 4.31, 6.7, 8.6
Combustible storage, 4.39
Commercial cooking, 9.10
Compliance alternatives, 34.5
Concealed spaces, 7.37
Concrete, anchorage to, 19.6
Concrete construction, 17.4
Concrete, durability requirements,
   19.2
Concrete-filled pipe columns, 19.12
Concrete, mixing, 19.4
Concrete, placing, 19.4
Concrete, structural plain, 19.6
Concrete, quality, 19.4
Constant attended location, 9.4
Construction, light-frame, 23.18
Construction documents, 1.4, 2.2,
   16.2
Construction joints, 19.5
Construction safeguards, 33.1
Contractor responsibility, 17.11
Controlled low-strength material, 2.2
Conveying systems, 30.3
Conveyors, 4.46
Corrosion resistance, 2.2
Courts, 12.3
Criminal facilities, 4.23
Covered mall buildings, 4.1

D

Dampers, fire, 7.35
Dampers, ceiling radiation, 7.37
Dampers, smoke, 7.34, 7.36
Dampproofing, 18.16

Dead load, 16.7
Decorative materials, 2.2, 8.6
Demolition, 33.1
Design requirements, 16.3
Design strength, 17.11
Detection systems, 9.14
Diaphragm, flexible, 16.2
Door, balanced, 10.8
Doors, 7.29, 10.10
Draft stopping, 7.39
Dressing rooms, 4.32
Driven pile foundation, 18.22
Dry cleaning, 4.48
Drying rooms, 4.53
Ducts, 7.33
Dwelling, 2.3
Dwelling unit, 2.3, 11.8

E

Earthquake loads, 16.25
Education, 3.5
Egress, 10.1
Egress width, 10.5
Embedded pipes, 19.5
Emergency alarms, 4.43
Emergency escape, 10.28, 10.33
Emergency operations, 30.2
Emergency power, 4.52
Engineered-wood products, 23.21
Equipment platform, 5.1, 5.5
Excavation, 18.4
Excess hazardous materials, 4.44
Exhaust ventilation, 4.52
Existing structure, 2.3
Exit, 10.5, 33.5
Exit access, 10.24
Exit-access doorways, 10.26
Exit, horizontal, 10.11, 10.30
Exit signs, 10.22
Expanded-vinyl wall covering, 8.1
Exterior exit ramps, 10.10
Exterior structural members, 7.27
Exterior-wall envelope, 14.2
Exterior walls, 4.17, 7.3

F

Fabric partitions, 16.2

Facilities, hazardous-materials, 4.40
Factory and Industrial, 3.5
Fiberboard, 23.2
Fill, 18.4
Fire barriers, 7.13
Fire classification, 15.3
Fire command center, 9.24
Fire dampers, 7.35
Fire escapes, 34.2
Fire extinguishers, portable, 9.14
Fire lane, 2.3
Fire partitions, 7.19
Fire resistance, 17.6
Fire resistance, calculated, 7.41
Fire-resistant joint systems, 7.24
Fire resistance, prescriptive, 7.41
Fire-resistance rating, 7.1
Fire-resistance separations, 4.4
Fire-retardant-treated lumber, 23.4
Fire shutters, 7.32
Fire test, 7.1
Fire walls, 7.10
Fire windows, 7.13
Flame spread, 8.1
Flame spread index, 8.1
Flammable liquids, application of, 4.53
Flashing, 14.4
Flexural length, 18.15
Floor joists, 23.20
Flood loads, 16.24
Floor framing, 6.5
Floor number signs, 10.20
Floors, 6.6, 10.13
Footings, 18.7
Formwork, 19.5
Foundations, 18.7
Foundation investigations, 18.1
Framing, floor, 6.5
Framing, roof, 6.5

G

Gas detection, 4.51
Gates, 10.15
Glass, general requirements for, 24.1
Glass loads, 24.2
Glazing, safety, 24.4

Grade floor opening, 2.3
Grade plane, 5.1
Grading, 18.4
Group A, 9.2, 9.15
Group A-1, 3.2
Group A-2, 3.2
Group A-3, 3.3
Group A-4, 3.3
Group A-5, 3.4
Group B, 9.15
Group E, 9.3, 9.16
Group F, 9.3, 9.16
Group F-1, 3.6
Group F-2, 3.8
Group H, 9.4, 9.16, 9.20
Group I, 9.4, 9.16
Group I-3, 3.11
Group I-4, 3.11
Group M, 9.4, 9.17
Group R, 9.5, 9.17
Group S, 9.5
Group S-1, 3.14
Guards, 10.23, 24.6
Gypsum construction, 25.4
Gypsum board, horizontal assemblies of, 25.1
Gypsum board materials, 25.2
Gypsum board, vertical assemblies of, 25.1
Gypsum board materials for showers, 25.6

H

Habitable space, 2.3
Handrails, 10.22, 24.6
Hangers, 4.35
Hardware, 10.14
Hazardous-materials facilities, 4.40
Hazardous uses, 3.8
Historic buildings, 34.3
Height, 5.2
Height, building, 5.2
Height, story, 5.2
Heliports, 4.37
Helistops, 4.37
High-rise buildings, 4.8
Historic buildings, 2.3

Hoistway enclosures, 30.1
Hoistway venting, 30.2
Horizontal assemblies, 7.21
Horizontal exit, 10.11, 10.30
Hydrogen cutoff rooms, 4.54

I

Illumination, means-of-egress, 10.5
In-situ load tests, 17.12
Insulation, foam-plastic, 26.1
Interior floor finish, 8.2, 8.4
Interior space dimensions, 12.4
Interior wall and ceiling finish, 8.2
Intumescent coatings, 17.8
Institutional uses, 3.9
Inspections, 1.5
Insulation, 15.8

J

Jalousies, 24.2
Jurisdiction, 2.3

L

Landings, 10.13
Lathing, 25.3
Lighting, 12.3
Light-frame construction, 2.3
Live loads, 16.9
Load-bearing values of soils, 18.6
Load-bearing wall, 2.6
Load combinations, 16.5
Log buildings, 23.3
Lot, 2.3
Lot line, 2.4
Lots, shared, 7.4

M

Machine rooms, 30.4
Mall buildings, covered, 4.1
Marquee, 2.4, 31.4
Masonry, AAC, 21.4
Masonry, ashlar, 21.15
Masonry construction, 17.5, 21.5
Masonry construction materials, 21.1
Masonry, empirical design of, 21.10
Masonry fireplaces, 21.17
Masonry, glass-unit, 21.5, 21.16

Masonry, rubble stone, 21.11
Masonry, quality of assurance of,
    21.7
Masonry walls, multiwythe, 21.12
Mastic coatings, 17.8
Materials, 1.2, 14.3, 15.5
Means of egress, accessible, 10.6,
    10.7
Means-of-egress illumination, 10.5
Medical facilities, 4.20
Membrane structures, 31.1
Mercantile, 3.12
Metal composite materials, 14.11
Mezzanine, 5.2, 5.3
Micropiles, 18.15, 18.23
Minimum slab provisions, 19.6
Mixed occupancies, 4.46, 5.9
Monitoring, 9.8
Mortar, Type N, 21.3
Mortar, Type S, 21.3
Mortar, thin-bed, 21.10
Motion-picture projection rooms,
    4.28
Motor fuel-dispensing facilities, 4.19
Motor-vehicle-related occupancies,
    4.14
Multiple-station alarm device, 9.7

N

Nosing, 10.14

O

Occupancy category, 16.2
Occupancies, 4.53
Occupancies, accessory, 5.11
Occupancies, change of, 34.3
Occupant load, 10.3
Occupiable space, 2.4
Open parking garages, 4.15
Opening protectives, 7.28
Organic coatings, 4.53
Outside walls, 7.6

P

Parapets, 7.9
Parking, 11.6
Parking garages, 4.15

Partitions, fire, 7.19
Passenger loading facilities, 11.6
Pedestrians, protection of, 33.2
Pedestrian walkway, 2.5, 31.3
Penetrations, 7.22
Penthouse, 15.8
Performance requirements, 14.1, 15.2
Permit, 2.4
Permits, 1.2
Person, 2.4
Piers, 18.15
Pier foundation, 18.15, 18.18, 18.24
Piles, 18.5
Pile foundation, 17.5, 18.15, 18.18
Pile foundation, driven, 18.22
Plaster, 7.40
Plaster, exposed aggregate, 25.8
Plaster, exterior, 25.7
Plaster, interior, 25.7
Plastering, 25.3
Plastic interior finish, 26.4
Plastic, light transmitting, 26.5
Plastic, light-transmitting glazing, 26.6
Plastic, light-transmitting interior signs
Plastic, light-transmitting roof panels, 26.8
Plastic, light-transmitting skylight glazing, 26.8
Plastic, light-transmitting wall panels, 26.6
Plastic signs, 4.7
Plastic trim, 26.4
Plastic veneer, 26.4
Platform construction, 4.32
Platform lifts, 11.14
Platforms, 4.29
Portable fire extinguishers, 9.14
Power-operated sliding doors, 4.26
Prescriptive fire resistance, 7.41
Public entrance, 11.5
Purlins, 23.21

R
Rain loads, 16.22
Ramps, 10.21

Ramps, exterior exit, 10.30
Registered design professional, 2.4
Reinforced gypsum concrete, 19.12
Repairs, 1.3, 34.1
Reroofing, 15.10
Residential, 3.12
Rescue, 10.33
Roof assembly, 15.4
Roof coverings, 15.5
Roof decks, 6.6, 15.2
Roof framing, 6.5
Roof vents, 4.31

S
Safety glazing, 24.4
Sanitary provisions, 33.2
Security, 4.7
Seismic design, 21.9
Seismic requirements, 23.24
Shaft enclosures, 7.14
Shafts, other types of, 7.17
Shared lots, 7.4
Shear-wall construction, 25.1
Shotcrete, 19.7
Signage, 11.15
Signs, exit, 10.22
Signs, floor number, 10.29
Signs, plastic, 4.7
Site class, 16.14
Site work, 33.2
Skylight unit, 2.4
Skylights, 7.22, 24.3
Skylights and sloped glazing, 2.4, 7.22
Sleeping unit, 2.4, 11.8
Sloped glazing, 24.3
Smoke control, 9.20
Smoke dampers, 7.34, 7.36
Smoke barriers, 7.20
Smoke-developed index, 8.2
Smokeproof enclosures, 10.29
Snow loads, 16.14
Soil investigations
Soil lateral loads, 16.22
Soils, 17.5
Soils, load-bearing values of, 18.6
Sound-insulating materials, 7.40

Sound transmission, 12.4
Special amusement buildings, 4.33
Special inspections, 17.2, 17.9, 18.5
Special occupancies, 11.11
Special provisions, 5.13
Sprinkler systems, automatic, 4.32,
    9.2, 9.6, 33.5
Stages and platforms, 4.29
Stage doors, 4.30
Stage exits, 4.32
Stairways, 10.17
Stairways, exterior, 10.30
Standpipe system, 9.10, 33.5
Steel, cold-formed, 22.4
Steel buildings, 17.3
Steel cable structures, 22.3
Steel identification, 22.1
Steel joists, 22.2
Steel protection, 22.1
Steel storage racks, 22.4
Stop-work order, 1.6
Storage, 3.13, 4.43
Story, 2.5
Story above grade plane, 2.5
Story height, 5.2
Structures, 15.8
Structural members, 7.26
Structural members, exterior, 7.27
Structural steel, 22.1
Surrounding materials, 12.5
Swimming-pool enclosure, 31.5

T

Temporary structures and uses, 1.4,
    31.3
Tent, 2.5
Termites, 23.8
Test standards, 17.12
Thermal materials, 7.40
Tower, radio, 31.4
Tower, television, 31.4
Townhouse, 2.5
Transfer openings, 7.33
Transport, 4.44
Trim, 8.2, 8.6
Trusses, 23.3
Turnstiles, 10.15

Type V construction, 9.8

U

Underlayment, 15.6
Underground buildings, 4.11
Unlimited-area buildings, 5.7
Unoccupied spaces, access to, 12.5
Utility and miscellaneous, 3.14
Utility services, 1.6
Use and occupancy classification,
    3.15

V

Vapor-permeable membrane, 2.5
Vapor retarder, 2.5
Vehicle barrier system, 16.2
Veneer, 14.3
Veneer, plastic, 26.4
Ventilation, 12.1
Ventilation systems, 4.18
Vertical openings, 4.26
Vinyl siding, 14.8
Violations, 1.6

W

Walkway, pedestrian, 2.5
Wall, load-bearing, 2.6
Walls, exterior, 4.17
Walls, fire, 7.10
Wall, outside, 7.6
Waterproofing, 18.16
Wind loads, 16.14
Window sills, 14.8
Windows, 7.9
Windows, fire, 7.13
Wood, decay of, 23.8
Wood, general construction
    requirements of, 23.4
Woods, general design requirements
    for, 23.9
Wood, minimum standards of, 23.1
Wood, quality of, 23.1

Y

Yard, 2.6
Yards, 12.3